The World Nuclear University Primer

Nuclear Energy

in the 21st Century

3rd Edition

Ian Hore-Lacy

World
Nuclear
University
Press

The World Nuclear University Primer
Nuclear Energy in the 21st Century, 3rd Edition
© 2012 World Nuclear Association. All rights reserved.

The first edition under this title was published by Elsevier and the World Nuclear University in 2006. The second edition was published by the World Nuclear University in 2010, reprinted 2011. Seven previous editions were published as *Nuclear Electricity* (1978-2003).

The front cover image was provided by Atmea (an Areva and Mitsubishi Heavy Industries joint venture company) and shows the ATMEA1 reactor.

The image on page 3 shows the bottom of the containment vessel being lowered into place at Sanmen unit 1 in December 2009. This is one of over 20 large reactors under construction in China, including modern western designs from Areva and Westinghouse.

Cover design by Richard Petrie
Book design by Brigita Praznik, Raf Damiaens and Richard Petrie

Printed in the United Kingdom by Clever Marketing

ISBN: 978-0-9550784-5-3

World Nuclear University Press
Carlton House, 22a St James's Square
London SW1Y 4JH, United Kingdom

www.world-nuclear.org

ABOUT THE AUTHOR

Ian Hore-Lacy, a former biology teacher, became General Manager of the Uranium Information Centre, Melbourne in 1995 and Head of Public Communications for the World Nuclear Association, based in London, in 2001. He has visited a number of nuclear reactors and fuel cycle facilities in several countries, including UK reprocessing plants, Sweden's waste facilities, US waste repositories and French enrichment and mixed oxide fuel fabrication plants.

He joined the mining industry as an environmental scientist in 1974 and gained some acquaintance with uranium mining. From 1988-93 he was Manager, Education and Environment with CRA Limited (now Rio Tinto) and has written several books on environmental and mining topics. His particular interests range from the technical to the ethical and theological aspects of mineral resources and their use, especially nuclear power. He has four adult children.

ACKNOWLEDGEMENTS

This text builds on seven editions of *Nuclear Electricity*, 1978-2003, published in Melbourne. The first edition under the current title was published by Elsevier and the World Nuclear University (WNU) in 2006, and the second by WNU in 2010 (reprinted 2011).

Section 3.7 on *Physics of a nuclear reactor* draws heavily on material written by Dr Alan Marks. Chapter 10 draws heavily on the introductory section of *Atomic Rise and Fall: the Australian Atomic Energy Commission, 1953-1987* by Clarence Hardy, Glen Haven Publishing (1999), and the Russian part of it was largely contributed by Judith Perera. In all cases material is used with permission.

Earlier editions of this book (*Nuclear Electricity*) owed their substance to Ron Hubery as co-author. Ron is a chemical engineer, now retired, who spent eight years working with the Australian Atomic Energy Commission (now the Australian Nuclear Science and Technology Organisation) on nuclear fuel cycles and reprocessing. He also worked at the uranium production centres of Rum Jungle and Mary Kathleen in Australia.

I am very grateful for the meticulous editing of this edition by Stephen Tarlton.

Ian Hore-Lacy

CONTENTS

FIGURES AND TABLES

FOREWORD

By Dr. Patrick Moore

Today our foremost energy challenge is to meet increasing needs without adding to our environmental and economic problems, notably air pollution and the cost of energy.

Though there is much talk of the need to severely limit greenhouse gas emissions, a significant reduction seems unlikely given our continued heavy reliance on fossil fuels. Yet nuclear energy offers the possibility of eventually replacing much of the fossil fuel used to generate electricity, and if battery-powered automobiles become widely used, much of the oil used to power transportation.

But environmental activists, notably Greenpeace and Friends of the Earth, continue to lobby against clean nuclear energy, and in favour of the failed Kyoto Treaty, offering unrealistic claims about replacing reliable base-load power with highly intermittent, unpredictable, and costly wind and solar energy. We can agree that renewable energies such as hydroelectric, biomass, and geothermal are part of the solution. But nuclear energy will prove to be even more effective in replacing fossil fuels and satisfying global demand for energy, and there is virtually no limit to the fuel supply for hundreds of years. The rigid and anti-scientific opposition to this proposition goes back to the mid-1980s when Greenpeace and much of the environmental movement made a sharp turn to the political left, adopting extreme agendas that abandoned science and logic in favour of emotion and sensationalism.

For the past two decades I have pursued the concept of sustainable development and sought to develop an environmental policy platform based on science, logic, and the recognition that seven billion people need to survive and prosper, every day of the year. Environmental policies that ignore science can actually result in increased risk to human health and the ecosystem. The zero-tolerance policy against nuclear energy that has been adopted by so many activist groups is a perfect example of this outcome. By scaring people into fearing atomic energy they virtually lock us into a future of increasing fossil fuel consumption.

Even though there have been three serious accidents at nuclear power plants during the 60 years they have been operating, nuclear energy is still one of the safest energy technologies we have invented. Not one person was killed by radiation at either Three Mile Island in the USA or at Fukushima in Japan, and according to the best experts there will be no discernable health effects from either incident. Chernobyl was an exception as the Russians designed a reactor that was inherently unsafe and will never be built again. Even so there were very few deaths – 56 according to the World Health Organization – compared with other major industrial accidents.

That is why I am pleased to commend this book, effectively a tenth edition of a comprehensive introduction to nuclear power, with a scientific basis and pitch. That is where I believe discussion and public debate on the question – and energy policies generally – needs to begin and remain based.

Nuclear energy can play a number of significant roles in improving the quality of our environment while at the same time providing abundant energy for a growing population. First, as mentioned above, it can replace coal and natural gas for electricity production. Coal-fired power plants alone produce about 30% of global CO_2 emissions. Under present scenarios, even with aggressive growth in renewable technologies, coal and natural gas consumption will continue to increase rather than decrease. The only available technology that can reverse this trend is nuclear energy.

France, for example, now obtains over 75% of its electricity from nuclear plants. Another 12% is hydroelectric therefore making France's electricity very low in fossil fuel use. Both Sweden and Switzerland, through a combination of nuclear and hydroelectric energy, provide a high standard of living with virtually no fossil fuel used for electricity production. If other countries had followed these three countries' example there would be far less fossil fuel used for power production than there is today.

Second, electricity from nuclear plants can be used to run ground-source heat pumps, also known as geothermal heat pumps, in all buildings. Buildings consume about 35% of the energy in an industrialized country, mostly using fossil fuels for heating and domestic hot water production, and electricity from the grid, often produced with fossil fuels. Heat pumps can provide heating, hot water, and air conditioning with no fossil fuels if the electricity is produced with nuclear, hydroelectric, or other renewables.

Third, nuclear energy can be used to desalinate seawater to provide water for drinking, industry, and irrigation. A growing population, shrinking aquifers, and increased irrigation demand all add up to the need to make our own fresh water in the future. Nuclear can provide the energy to do it without causing pollution or greenhouse gas emissions.

Fourth, high-temperature nuclear reactors can be used to produce hydrogen for stationary fuel cells that replace natural gas and petroleum as a source of hydrogen in the petrochemical and coal-to-liquid fuel industries. A nuclear plant can produce sufficient heat to split water into hydrogen and oxygen thermally. This is much more efficient than using electricity to split water. There are a lot of technical hurdles to overcome, and the hydrogen economy may still be years away, but there is no other alternative to using fossil fuels for hydrogen production in the offing.

We will continue to use fossil fuels, hopefully at reduced levels, far into the future. As conventional supplies of oil diminish we will turn to the vast shale gas, shale oil, and oil sand deposits. This is already a growing industry in northern Canada where the oil sands contain as much proven supply as Saudi Arabia. But the oil costs more because it must be separated from the sand. This is done by burning large volumes of natural gas to make steam; then basically steam-cleaning the sand to get the oil. By using one fossil fuel to obtain another there is even more greenhouse gas emissions than from using conventional oil supplies. One solution to this would be to use nuclear energy to make the steam, and electricity, to run these oil sand and shale oil projects. This would substantially reduce greenhouse gas emissions and air pollution.

There are 433 nuclear reactors operable in 30 countries producing 13% of the world's electricity. There are over 60 reactors under construction at this printing. The production of nuclear energy could be doubled or tripled if the political will were brought to bear on the issue of reducing fossil fuel consumption. I believe that the environment would benefit from moving in this direction. Let's hope the future takes us there.

Co-founder of Greenpeace, Dr. Patrick Moore is Chairman and Chief Scientist of
Greenspirit Strategies Ltd. in Vancouver, Canada.
www.greenspiritstrategies.com

INTRODUCTION

There is a rapidly-increasing world demand for energy, and especially for electricity. Much of the electric demand is for continuous, reliable supply on a large scale, which generally only fossil fuels and nuclear power can meet.

The fuel for nuclear power to make electricity is uranium, and uranium's only substantial non-weapons use is to power nuclear reactors. There are over 850 nuclear reactors operating today around the world. These include:

- About 240 small reactors, used for research and for producing isotopes for medicine and industry in 56 countries.
- About 180 small reactors powering about 150 ships, mostly submarines.
- Some 433 larger reactors generating electricity in 30 countries.

Practically all of the uranium produced today goes into electricity production (though a significant small proportion is used for producing radioisotopes). In particular, uranium is generally used for base-load electricity. Here it competes with coal, and in recent years, natural gas.

Over the last 55 years nuclear energy has become a major source of the world's electricity. It now provides over 13% of the world's total. It has the potential to contribute much more, especially if greenhouse concerns lead to a change in the relative economic advantage of nuclear electricity.

Some decades away there is an emerging prospect of the 'hydrogen economy', with much transport running on hydrogen. Just as nuclear power now produces electricity as an energy carrier, it is likely to produce much of the hydrogen, another energy carrier.

The uranium and nuclear power debate today is about options for producing electricity. None of those options are without some risk or side effects.

Since the first edition of this book in 1978 – as *Nuclear Electricity* – many of the inflated expectations of alternative energy sources have been shown to be unrealistic (as have some of those for nuclear energy). However, it is important that this return to reality does not lead to their neglect – such alternatives should continue to be developed, and applied where they are appropriate. In particular, a great deal can be achieved by matching the location, scale and thermodynamic character of energy sources to particular energy needs. Such action should be a higher priority than merely expanding capacity to supply high-grade electrical energy where for example only low-grade heat is required, or using versatile gas to generate electricity on a large scale simply because the plant is cheaply and quickly built.

In physical terms, the prime attribute of nuclear power is its energy density. In a nuclear reactor there is a concentration of energy produced that is orders of magnitude greater than with any other commercial technology.

The present and future roles of nuclear power are not limited to electricity, and hence the expanded scope of this book beyond *Nuclear Electricity*. The large potential for nuclear heat in process applications and later to make hydrogen to fuel motor vehicles is perhaps the chief interest today. But this book also encompasses the use of nuclear energy in the production of potable water through desalination, marine propulsion, space exploration, and the production of radioisotopes.

When the question of utilising nuclear energy arises, there are those who wish somehow to put the genie back in the bottle and to return to some pre-nuclear innocence. The situation in Europe is instructive: France gets over 75% of its electricity from nuclear power. It is the world's largest electricity exporter, and gains some €3 billion per year from those exports. Next door is Italy, a major industrial country without any operating nuclear power plants. It is the world's largest net importer of electricity, and most of that comes ultimately from France. A few countries cling to nuclear phase-out policies which are patently unrealistic.

I anticipate that my grandchildren's generation will come to look upon weapons as simply an initial aberration of the nuclear age, rather than a major characteristic of it. Arguably the same is true of the bronze and iron ages, where weapons provided incentive for technological development which then became applied very widely.

Certainly, as Figure 1 in Chapter 1 graphically shows, we cannot indefinitely depend on fossil fuels as fully as we do today.

The book

Considerable effort has been made to include as much up-to-date and pertinent information as possible on generating electricity from nuclear energy, and on other uses of nuclear power. The figures quoted are conservative, and generalisations are intended to withstand rigorous scrutiny. The reader will not see many of the frequently repeated assertions from supporters or opponents of nuclear energy.

Since the first edition of *Nuclear Electricity*, the intention has been to get behind the controversies and selective arguments, and present facts about energy demand and how it is met, in part, by nuclear power. Every form of energy production and conversion has an effect on the environment and carries risks. Nuclear energy has its challenges but these are frequently misunderstood and often misrepresented. Nuclear energy remains a safe, reliable, clean, and generally economic source of electricity with minimal impact on the environment. But many people do not see it that way, especially since the March 2011 Fukushima accident in Japan where in the context of a natural disaster claiming 19,000 lives, no lives were lost as a result of the nuclear accident. However, at the time of writing, the many evacuees are suffering a lot, mostly for no evident good reason.

This edition comes out at a time when environmental concern focused on tangible indicators of pollution and global warming stands in ever-starker contrast to Romantic environmentalism, which is driven by mistrust of science and technology, and which stigmatises nuclear power. Increasing evidence of the contribution to global warming from burning fossil fuels is countered by fear-mongering often based on the 1986 Chernobyl disaster and now compounded by the Fukushima accident.

The introduction to the first edition of this book in the 1970s expressed the opinion that if more effort were put into improving the safety and effectiveness of commercial nuclear power, and correspondingly less into ideological battles with those who wished it had never been invented, then the world would be much better off. With Chernobyl nearly a quarter of a century behind us and the great improvements to safety in those plants which most needed it, plus the welcome recycling of military uranium into making electricity, it seems that we are now closer to that state of affairs.

As John Ritch, President of the World Nuclear University, pointed out in opening the sixth annual WNU Summer Institute: "Between now and 2050, as world population swells from 6.8 billion toward 9 billion, humankind will consume more energy than the combined total used in all previous history. Under present patterns of energy use, the consequences will prove calamitous. The resulting pollution will damage or ruin the health of tens and likely hundreds of millions of citizens, mainly in the developing world. Far worse, the intensifying concentration of greenhouse gases will take us past a point of no return as we hurtle toward climate catastrophe. Today the world economy is producing greenhouse emissions at the rate of 30 billion tonnes per year – nearly 1,000 tonnes per second." He made it clear that "present patterns of energy use", if continued, would have serious consequences. This book provides some detail of an alternative.

Further information

All the matters covered in this book can be explored in more detail. One convenient way of doing so is by accessing the Public Information Service section of the World Nuclear Association website:

www.world-nuclear.org

It has some 150 Information Papers on specific topics, as well as links to other websites with reliable information.

Nuclear Energy
in the 21st Century

1. Energy use

1.1 SOURCES OF ENERGY

All energy is derived ultimately from the elemental matter which comprises both the Sun and the Earth, formed in supernovae over six billion years ago. From the Sun we have both energy trapped in fossil fuels and the main contemporary renewable sources. From the elemental substance of the Earth we have uranium and geological heat.

The Sun warms our planet, and provides the light required for plants to grow. In past geological ages, the Sun provided the same kind of energy inputs. Its energy was incorporated into the particular plant and animal life (biomass) from which were derived today's coal, oil and natural gas deposits – the all-important fossil fuels on which our civilisation depends.

The only other ultimate energy source in the Earth is from the atoms of particular elements formed before the Solar System itself. These are found today in the Earth's crust[1] and mantle. The amount of energy in an atom is dependent on the size of the atom: the minimum amount of energy per unit mass is contained within the medium-sized atoms (such as carbon and oxygen), while the greatest amount is contained in small atoms (such as hydrogen) or large atoms (such as uranium). Energy can therefore be released by combining small atoms to produce larger ones (fusion) or by splitting large atoms to produce medium-sized atoms (fission). The tapping of this energy by nuclear fission or by nuclear fusion is one of the most important and contentious human achievements in history.

1.2 SUSTAINABILITY OF ENERGY

Much has been written since the early 1970s about the impending 'world energy crisis', which was initially perceived as a crisis due to limited oil supplies. Today it is more a geopolitical crisis due to the location of supplies of oil and gas resources relative to demand for them. But finite supplies are still a factor, and Figure 1 suggests the vital importance of conserving fossil fuel resources for future generations, and the importance of sustainability.

While since the early 1970s the pressure has been to conserve crude oil supplies, in future it will increasingly be to reduce burning of all fossil fuels. Today, climate change concerns drive this trend strongly. It is likely that coal will take over some of the roles of oil today, especially as a chemical feedstock. Sustainability of energy relates both to adequacy of supplies where energy is needed, and the environmental effects of its use.

The importance of energy conservation is obvious, even in areas where so far fuels have been relatively cheap, and limiting carbon emissions lends emphasis to this. The levelling-out of overall energy demand in developed countries in recent decades is a result of increased energy efficiency. However, in developing countries growth in energy demand from a low starting point continually increases the pressures on resources worldwide, despite conservation initiatives (see Table 1).

Many people in developing nations aspire to the standard of living, mobility, agricultural productivity and industrialisation characteristic of the developed countries. Fulfilling these hopes depends on the availability of abundant energy. Growth of the world's population from the present level of 7 billion people to a projected 8.6 billion in 2035, mostly in today's developing nations, increases the challenge.

Figure 1. Consumption of fossil fuels

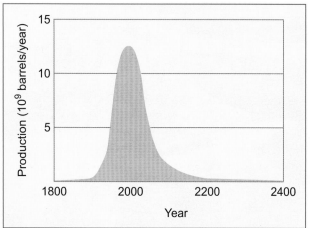

Hubbert curve scaled to peak at about 2020

---

[1] Uranium appears to have been formed in super novae some 6.5 billion years ago, and though not common in the Solar System, has been concentrated in the Earth's crust at an average of c 1.4 ppm. Heat from the radioactive decay of this uranium today drives the convection processes in the Earth's mantle, and is vital to life.

Table 1. Growing electricity demand

	1990 (TWh)	2009 (TWh)	Increase
OECD	6,593	9,193	39%
Non-OECD	3,492	8,024	130%
E. Europe/Eurasia	1,585	1,280	-19%
Africa	263	532	102%
Latin America	404	816	102%
Asia (ex China)	490	1,533	213%
China	559	3,263	484%
Middle East	190	600	216%
World	10,084	17,217	71%

Figures in billion kilowatt hours, kWh, or terawatt hours, TWh
Source: OECD/IEA, World Energy Outlook 2011, Table 5.1.

1.3 ENERGY DEMAND

In industrialised countries energy demand is in three major sectors: domestic and commerce; industry and agriculture; and transport.

In many countries, these each account for about one-third of the energy demand, although the size of domestic demand depends very much on climate. In Australia domestic demand is relatively small, whereas in Canada for example, it is relatively large because of the cold climate. More specifically it is possible to identify demand for particular purposes within these sectors, such as:

• Low temperature heat (up to 110°C) for water and space heating in homes and industry.
• High temperature heat (over 110°C) for industrial processes.
• Lighting.
• Motive power for factories, appliances and some public transport.
• Mobile transport for public and private use.

For some of these purposes there is a significant demand for energy in the form of electricity. Worldwide, electricity demand is increasing very rapidly, as illustrated in Table 1 above, and Figure 4 on page 8. This is discussed further in Section 2.1.

1.4 ENERGY SUPPLY

On the supply side, there are a number of primary energy sources available (see Figure 3 on page 7). Derived from these primary sources are several secondary energy sources or carriers. These include, for example:

• Electricity – can be generated from many primary sources.
• Hydrogen – mainly from natural gas.
• Alcohols – from wood and other plant material.
• Oil and gas – can be manufactured from coal.

At this stage, only electricity is of major importance as a secondary source, but hydrogen is expected to become significant in the future as a replacement for oil products (see Chapter 7).

Much energy demand can be met by more than one kind of energy supply. For instance, low temperature heat can be produced from any of the fossil fuels directly, from electricity, or from the Sun's radiant energy. Other demands such as mobile transport need to be supplied by portable and energy-dense fuels such as those derived from oil or gas, or from electricity. In the future, hydrogen is expected to become important in this role, but not for some years.

Both economic practicality and ethical considerations mean that versatile, energy-dense, easily portable energy sources such as oil and its derivatives should not be used where other, more abundant fuels can be substituted. Different energy sources yield different amounts of energy per unit mass or volume, as shown in Table 2 at the end of this chapter.

Primary energy resources in different countries vary enormously. There are great differences in natural endowment and this makes clear the importance of trade in energy, as indicated in Figure 2.

1.5 CHANGES IN ENERGY DEMAND AND SUPPLY

The uneven world distribution of energy resources means that as energy consumption rises, international trade in energy must increase. Energy-poor countries find themselves dependent on supplies from energy-rich countries as Figure 2 illustrates. Because of the fundamental importance of energy in the industrial economy, importing countries are vulnerable politically and economically. Energy trade between regions is projected to double by 2030, and most will continue to be in the form of oil.

The best illustration of this vulnerability is the changing position of oil. Until the early 1970s, many countries had come to depend on oil because of its relatively low cost, and world oil production tripled between 1960 and 1973. But this suddenly changed as prices rose four-fold, and then there was a further 'oil crisis' in 1979. As a result, world oil consumption in 1986 was the same as that in 1973, despite a substantial rise in total primary energy consumption. Forecasts in 1972 had generally predicted a doubling of oil use in ten years.

Problems of oil prices and supply in the 1970s brought about rapid changes in the production and use of other primary energy resources:

- Coal production and international trade in coal increased to substitute for some oil use.
- Nuclear power for electricity generation was adopted or examined more closely by energy-deficient countries.
- Most countries looked more closely at adopting measures to restrain energy consumption, and this focus continues.
- Renewable energy sources were studied seriously (in some cases for the first time) to determine whether and where they could be used economically.

Japan for example, has few indigenous energy resources and little untapped hydro-electric potential. It suddenly found that escalating oil imports to supply three-quarters of its total energy needs were not sustainable. Even the USA, originally self-sufficient in oil, found it difficult to pay for enough imported oil to offset declining domestic production. The USA has also been the world's largest

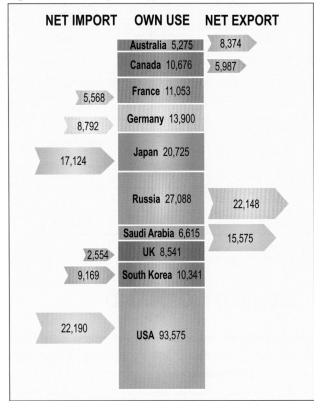

Figure 2. Primary energy in 2010 (PJ)

Area of colour is indicative to amount total primary energy supply (TPES). Russia & Saudi Arabia: 2009 figures
Source: OECD/IEA, Energy Balances of Non-OECD Countries – 2011 Edition; OECD/IEA, Energy Balances of OECD Countries – 2011.

importer of natural gas, and security of energy supply is a major factor in its foreign policy.

The thrust of these changes has continued into the new century. Throughout the world it was found possible to use significantly less energy per unit of economic activity. The use of oil for electricity production was greatly reduced and the use of natural gas increased. More recently the shipment of gas as liquefied natural gas (LNG) has enabled even greater use. Wind is now the fastest-growing source of electricity.

Continuing a trend predating the oil crisis, the demand for primary energy per unit of gross domestic product (*i.e.* energy intensity) has shown a significant decline (1.3% per year) in OECD[2] countries and this is expected to be the case also in developing countries in the future. However, at the same time the electricity consumption per unit of gross domestic product has been growing steadily, reflecting a strong increase in the proportion of electricity used in all countries.

---

[2] Organization for Economic Co-operation and Development

Figure 3. World primary energy demand

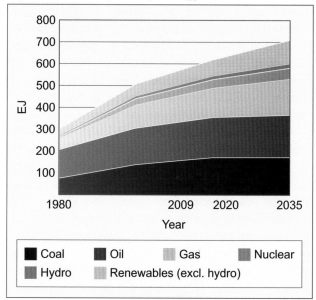

Source: OECD/IEA World Energy Outlook 2011, 'New Policies' scenario

Electricity is uniquely useful for driving machinery and for lighting in both industry and homes. However, it is also used for heating and in other ways for which alternatives are readily available. It can be argued that in view of the relatively low efficiency of energy conversion to electricity (typically around 35%), alternatives such as natural gas should be used wherever possible for heating (at double the efficiency)[3]. Conversely, it can be argued that uranium and coal resources are large relative to gas resources, that the most abundant primary fuel should be applied wherever possible, and that hence electricity use for heating (at almost 100% end use efficiency) is desirable if the electricity is produced from coal or nuclear power despite a much higher consumption of primary fuel. In relation to coal, this argument is heavily qualified due to the one kilogram per kilowatt-hour carbon dioxide (CO_2) emissions, however.

> The role of electricity is increasing because it is an extremely versatile energy source which can be generated from a wide range of fuels and can easily be reticulated to the point of use. Electricity generation uses some 40% of the world's total primary energy supply.

In one sense the Sun is the world's most abundant energy source and the desirability of applying it more widely to direct heating and even for generation of electricity on an increasing scale hardly needs emphasis. Meanwhile wind is increasingly harnessed for electricity. Questions concerned with the production of electricity from renewable sources are discussed in Section 2.5.

In the following chapters electricity demand, use and generation are the focus of discussion. In particular they discuss the use of nuclear energy to generate electricity. The main nuclear fuel concerned is uranium, a metal which at present has virtually no other civil uses.

1.6 FUTURE ENERGY DEMAND AND SUPPLY

Where will we obtain our future energy needs? There are a number of uncertainties:

- Oil production peaked in 1979 and did not return to that level until 1994. Prices depend significantly on political factors, making them volatile and generally high.
- Natural gas production, while increasing rapidly now due to its recovery from shales, is likely to approach its peak in many countries in the next couple of decades.
- Underground coal is costly to mine, and all coal use gives rise to concern about its effect on climate change.
- There is limited scope for utilising renewable energy resources; in the case of wind and solar due to their intermittency and low energy density.
- Further scope for energy conservation is limited without radical changes in lifestyle in developed countries, and is low in developing countries as per capita use is low.

Oil production platform in Bass Strait, Australia

---

[3] Considering the whole sequence from production to end use, the efficiency of gas and oil for heating is often about 40-45%. For modern high efficiency gas furnaces, the value increases to about 70%, but overall depends on distance from the gas sources.

Figure 4. World electricity consumption by region

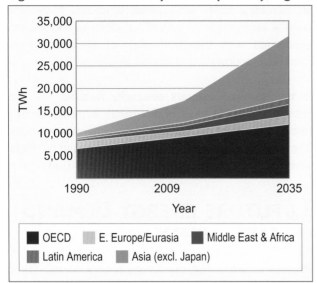

Source: OECD/IEA World Energy Outlook 2011, 'New Policies' scenario

Since the oil shocks of the 1970s, many industrialised nations set out to reduce dependence on oil and develop other strategies including greater use of nuclear energy. Looking ahead, it is not the industrialised countries which will dominate the scene – China has already overtaken the USA as the world's largest energy consumer, and by 2035 it is expected to use 70% more than the USA. It already accounts for almost half of world coal consumption. Energy demand in India was higher than in Russia in 2009, making it the world's third-largest energy consumer, and its energy demand is expected to more than double by 2035.

As economic growth occurs in most nations, and the world population grows towards 8.6 billion in 2035, increased energy demand is an inescapable part of this. Successive editions of the OECD/IEA *World Energy Outlook* examine the matter, and the *New Policies* scenario of the 2011 edition informs the comments here.

Fossil fuels will still account for 75% of primary energy consumption in 2035, compared with 81% now, and natural gas will increase its share.

Today, oil demand continues to increase, while available resources decline – production has exceeded discoveries since the 1980s, and despite very intensive exploration effort, consumption is now twice the rate of discovery. Most conventional reserves are in geopolitically uncertain parts of the world and are difficult to access. A lot of newly-discovered oil requires greater effort in both recovery and refining. Other reserves such as tar sands pose major problems to develop on any large scale.

Natural gas is less constrained, but again most conventional reserves are located in geopolitically uncertain areas and transport becomes a major problem. Moving it as liquefied natural gas (LNG) consumes around 30% of it. Shale gas is more widely distributed, but extracting it is problematic. Most of the projected increase in OECD demand for gas to 2035 comes from the power sector. Worldwide gas demand is projected to reach that for coal by this time, with 80% of the increase in demand coming from non-OECD countries.

Much of the projected increase in demand for coal to 2035 is for the power sector. Coal remains very abundant and is often available close to where it is needed – most coal is used in the country in which it is produced. (In 2010, international trade in coal reached 1083 million tonnes, accounting for about 16% of total coal consumed.) It is economically attractive to use on a large scale, but delivers the greatest contribution to greenhouse gas emissions of any fossil fuel. Carbon capture and storage (CCS) technology could boost the long-term prospects for coal, but that is unlikely to play a major role before 2035.

Nuclear output is expected to increase some 70% to 2035, maintaining its share of global electricity generation, thanks largely to China.

Much of the increase in electricity generation to 2035 comes from non-hydro renewables, which are expected to increase their share from 3% of the total to 15%. However, they cannot meet the extent of the demand. Cost, and the diffuse and intermittent nature of these sources limit their potential.

> **Electricity demand is growing much faster than overall energy demand.**

One-third of the world's population does not have access to electricity supply, and a further third does not enjoy reliable supply. There is a huge need to address these shortcomings and expectations, at the same time as implementing sustainable development principles and reducing poverty. Providing affordable power is a major part of the development solution.

There is also a rapidly increasing demand for potable water in many developing areas (for example, North Africa and the Arab Gulf States) that must be satisfied by desalinating facilities (see Section 7.4); this will further increase energy demand.

Energy consumption in OECD countries is forecast to increase only marginally to 2035, while that in developing countries with accelerating urbanisation and industrialisation is expected to grow very much faster. And where world primary energy demand is expected to increase 40% from 2009 to 2035, electricity demand is expected to grow 84% in that period, with China and India accounting for more than half of that increase. World electricity demand is projected to be almost 32,000 TWh (billion kWh) in 2035, compared with 17,200 billion kWh in 2009.

Since the 1970s, economic factors have constrained energy demand and have resulted in unprecedented increases in energy efficiency in industry and transport, at least in the OECD countries. The future scope for energy conservation depends on the sector involved. Where energy is a significant input to industrial processes or to transport, or an obvious cost to consumers such as with motor vehicles, major steps have already been taken to increase efficiency and hence lower costs. But where energy costs are relatively less significant, such as in commercial and residential buildings, it is likely that much scope for improvement remains.

Energy conservation is very difficult to project. To continue to be effective, it requires a present response to future prospects of higher energy costs. It demands an attitude to energy use and lifestyle which is increasingly conservation-oriented, so that the rate of increase in overall energy consumption remains depressed after the initial easy fixes have been achieved. Despite popular acceptance of environmental ideas, there is little evidence of such an attitude taking precedence over comfort and amenity anywhere in the world.

Table 2. Heat values and carbon coefficients of various fuels

	Heat value		% carbon	CO_2
Hydrogen	121	MJ/kg	0	0
Petrol/gasoline	44-46	MJ/kg		
	32	MJ/L		
Diesel fuel	45	MJ/kg		
	39	MJ/L		
Crude oil	42-44	MJ/kg	89	70-73 g/MJ
	37-39	MJ/L		
Methanol	20	MJ/kg	37	
	18	MJ/L		
Liquefied Petroleum Gas (LPG)	49	MJ/kg	81	59 g/MJ
Natural gas (UK, USA, Australia)	38-39	MJ/m³	76	51 g/MJ
(Canada)	37	MJ/m³		
(Russia)	34	MJ/m³		
as LNG (Australia)	55	MJ/kg		
Hard black coal (IEA definition)	>23.9	MJ/kg		
(Australia & Canada)	c 25.5	MJ/kg	67	90 g/MJ
Sub-bituminous coal (IEA definition)	17.4-23.9	MJ/kg		
(Australia & Canada)	c 18	MJ/kg		
Lignite/brown coal (IEA definition)	<17.4	MJ/kg		
(Australia, electricity)	c 10	MJ/kg	25	1.25 kg/kWh
Firewood (dry)	16	MJ/kg	42	94 g/MJ
Natural uranium*				
in LWR (normal reactor)	500	GJ/kg	0	0
in LWR with U & Pu recycle	650	GJ/kg	0	0
in FNR	28,000	GJ/kg	0	0
Uranium enriched to 3.5%, in LWR	3900	GJ/kg	0	0

NB natural uranium normally needs to be enriched to be used, the above figures are for it as mined
Uranium figures are based on 45,000 MWd/t burn-up of 3.5% enriched U in LWR
MJ = 10⁶ Joule, GJ = 10⁹ J; % carbon is by mass; mass CO_2 = 3.667 mass C
MJ to kWh @ 33% efficiency: x 0.0926
One tonne of oil equivalent (toe) is equal to 41.868 GJ
Sources: OECD/IEA Electricity Information 2008, for coal; Australian Energy Consumption and Production, historical trends and projections, ABARE Research Report 1999.

2. Electricity today and tomorrow

2.1 ELECTRICITY DEMAND

Electricity demand in an industrial society arises from a number of sources, including:

- Industry
    Some running on a 24-hour basis;
    Some working 8-10 hours only on weekdays.
- Commerce
    Most working 10-15 hours per day.
- Transport
    Public transport running during day and evening;
    Electric cars charging mostly off-peak overnight (in near future).
- Domestic
    Heating mostly during day and evening, seasonal;
    Cooling, seasonal;
    Cooking mostly in evening;
    Other household appliances;
    Water and space heating, off-peak overnight (in some systems).

It is clear from the above why electricity demand fluctuates throughout every 24-hour period as well as through the week, and also seasonally. It also varies from place to place and from country to country depending on the mix of demand, the climate, and other factors. Weekday load curves for an electricity system in a temperate climate are shown in Figure 5. From this it can be seen that there is a base-load of about 60% of the maximum load for a weekday. These load curves are typical for developed countries.

> **The base-load demand for continuous, reliable supply of electricity on a large scale is the key factor in any system. The main investment of any electric utility is to meet that kind of demand.**

As well as the daily and weekly variations in demand there are gradual changes occurring in the pattern of electricity demand from year to year. In projecting demand patterns a decade or more into the future, planners must take note of such factors as:

- The changing pattern of seasonal peak demands; for example as summer air conditioning becomes more common.
- The impact of increased electrification of public transport.
- The imminent electrification of private transport, with battery charging mostly overnight.
- The effect on supply systems of increasing use of solar water heating with electrical boosting during periods of adverse weather.
- The effect of incentives (e.g. through smart meters) to increase off-peak electricity demand and minimise peak demand, especially for water and space heating.
- The practical effect of energy conservation measures such as insulation and more energy-efficient building and appliance design.
- The role of renewable energy sources providing electricity when they can, and political coercion on utilities to buy or supply that electricity preferentially despite higher cost than other sources.
- Any increase in other dispersed electricity generation.
- Industry needs and how they are changing in response to economic factors.
- Improvements in the ability to transmit electricity long distances; e.g. 50 years ago, 600 km was the maximum distance for efficient transmission; in the 1960s, new technologies enabled transmission over 2000 km; and today it is greater still.

Looking further ahead, there is major scope for the use of base-load electricity to charge the batteries for personal motor vehicles. In the last few years, the popularity of hybrid cars such as the Toyota Prius has put us within reach of practical electrical motoring for many people. This development has been enabled by the advent of much more efficient battery technology[1]. While these vehicles in themselves make no difference to energy demand, a further increase in battery capacity with overnight charging from mains power will transform

Figure 5. Load curves for a typical electricity grid

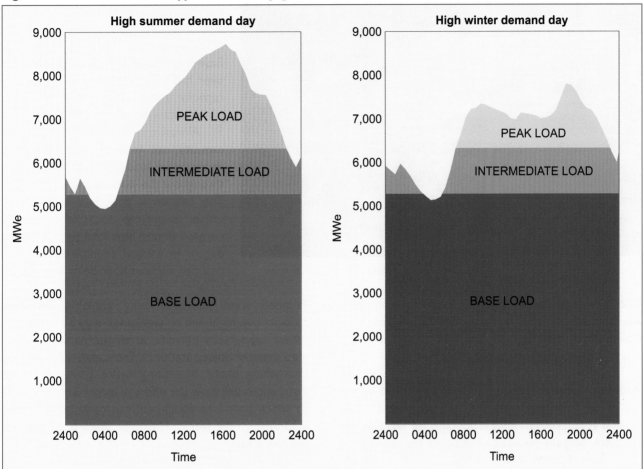

Summer and winter weekday load curves, midnight to midnight, showing the relative contributions of base-, intermediate- and peak-load plant duty. The shape of such a curve will vary markedly according to the kind of demand. Here, the peaks reflect domestic demand related to a normal working day, with household electric hot water systems evident overnight.
Note that the base-load here is about 5100 MWe, and while total capacity must allow for nearly double this, most of the difference can be supplied by large intermediate-load gas-fired plant or to some extent by adjusting the output of the base-load plant. The peak loads are typically supplied by hydro and gas turbines. In the wholesale electricity market, power stations bid into the market and compete for their energy to be despatched, so economic factors tend to determine the sources of supply at any particular moment.
Source: VENcorp (Australian Energy Market Operator)

the scene, and provide electromobility. There will be less reliance on any on-board internal combustion motor and more reliance on base-load power, mostly off-peak.

Some of the factors listed above will affect total electricity consumption, while others will influence the relative importance of base-load demand. For instance, off-peak charging in the load curves of Figure 5 would increase the base-load contribution by 35-40% (see Figure 5A on page 13), thereby reducing the average cost of electricity for all users. Production economics mean that as much of the electricity as possible should be supplied from base-load generating plant, since that from peak-load plant is usually significantly more expensive[2].

2.2 ELECTRICITY SUPPLY

Because of the large fluctuations in demand over the course of the day, it is normal to have several types of power stations broadly categorised as base-load, intermediate-load and peak-load stations.

The base-load stations are usually steam-driven and are most economical if run more or less continuously at near rated power output. Coal and nuclear power are the main energy sources used.

Intermediate-load and peak-load stations must be capable of being brought on line and shut down fairly

---

[1] The Toyota *Prius* nickel metal hydride battery in 2005 could deliver 21 kW from a mass of 45 kg. It held 6.6 amp hours at 201 volts (1.3 kWh). Plug-in hybrids will need at least 10 kWh on board.
[2] Furthermore, government policies in many countries mandate preferential input from any renewable generating capacity linked to the system, often regardless of cost.

China's electricity demand is expanding rapidly (photo: China Light & Power)

therefore normally about 20% more than the maximum load in a system, providing a reserve.

Base-load plants are likely to make up over half of a system's total generating capacity, and produce more than 85% of the total electrical energy (see Figure 5). Almost one-third of such a system's capacity can broadly be classified as intermediate-load plant, supplying power throughout the working day and evening. The balance is peak-load in the strict sense, supplying short-term energy demand during high load periods of the day or in unusual conditions, and with cost per kilowatt hour being less critical.

quickly once or twice daily. A variety of technologies are used for intermediate- and peak-load generation, including gas turbines, gas- and oil-fired steam boilers and hydro-electric generation.

Peak-load equipment tends to be characterised by low capital cost, and its relatively high fuel cost (unless hydro) is not a great problem since it operates for less than half the time, perhaps 60 out of 168 hours per week.

Base-load plant is designed to minimise fuel cost, hence usually fuelled by coal or uranium, and the relatively high capital cost can be written off over the large amounts of electricity produced continuously.

The capital cost of peak-load equipment such as gas turbines is about half that of base-load coal-fired plant, and in addition it can be installed much more quickly. However, the fuel cost is relatively high compared with coal in a base-load station, per unit of power generated. Modern combined cycle gas turbine facilities, which have efficiencies substantially greater than that of coal-fired plants, reduce the difference and in places have been used to supply base-load demand.

Lowest overall power costs to the consumer are obtained when the peak-load increment is very small and a steady base-load utilises most of the available generating capacity fairly constantly. Any practical system has to allow for some of the plant being unserviceable or under maintenance for part of the time. Installed capacity is

Pumped water storage, using available base-load capacity overnight and on weekends, may be developed where topography permits, as an alternative to peak-load thermal power stations[3]. The capital cost may be low where there is existing hydro plant, and such installations will have the effect of increasing the extent to which base-load equipment can contribute to total load through the week. However, there is a significant efficiency loss.

In future, the base-load contribution to supply is likely to be increased by using the surplus power in non-peak periods to charge electric vehicle batteries. Eventually it may also be used to make hydrogen, either for peak power generation or for transport fuel – see Section 7.2 for more detail. If we superimpose that overnight charging scenario on Figure 5 it looks like Figure 5A.

A further means of increasing the role of base-load plants is enabling them to follow the load to some extent, by varying the output. France's EDF has developed means of daily and weekly load-following across its fleet of nuclear plants.

As in other industries, there are economies of scale. Larger steam units result in reduced capital cost per kilowatt capacity, especially for base-load equipment. This means that location is sometimes determined as much by the supply of cooling water as by the fuel source. However, large power station units require a

[3] See also later part of Section 2.5. The 3000 MWe South Ukraine nuclear power plant is coupled with the Tashlyk 300 MWe pumped storage facility, producing 175 million kWh/yr.

Figure 5A. The effect of overnight charging of plug-in hybrid electric vehicles (PHEV) and electric vehicles (EV) on Figure 5

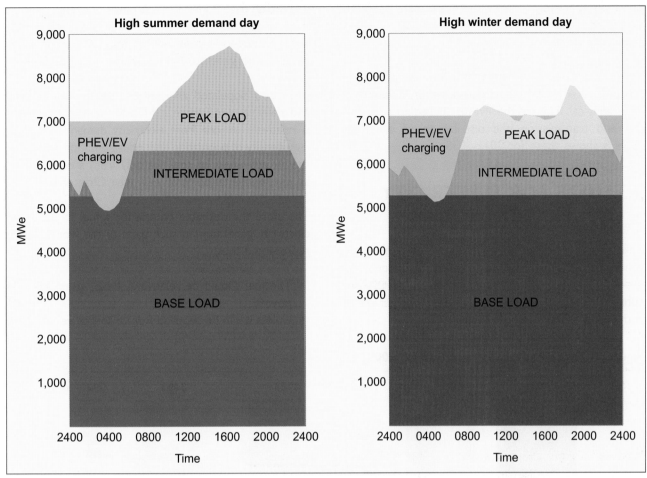

The UK Department of Transport has estimated that if the UK switched to electric vehicles, electricity demand would rise about 16%. The US Electric Power Research Institute modelled 60% of US vehicle use being electric, and found a 9% increase in electricity demand. As can be seen from the graphs above, this need not increase the system's peak capacity if most charging is off-peak, thereby greatly increasing the proportion of total generating capacity supplied by base-load plant. A study for the US Department of Energy in 2006 found that the idle off-peak grid capacity in the USA would be sufficient to power 84% of all vehicles in the USA if they all were immediately replaced with electric vehicles. In the above diagrams, assuming significant move to electric cars, the base-load demand is increased by about 35%.

Table 3. Electric capacity additions and required investment 2011-2035

	Capacity addition, GWe	Power generation investment, $ billion	Transmission & distribution investment, $ billion
North America	880	1,738	1,271
Europe	938	1,976	915
East Europe, Eurasia	331	588	442
Asia	2,893	4,106	3,486
Other	854	1,383	978
World total	**5,896**	**9,791**	**7,092**

Source: IEA World Energy Outlook 2011, 'New Policies' scenario

substantial electrical transmission grid and significant-sized overall generating system[4] to enable them to be operated effectively. Hence there are many situations where the economic virtues of small-scale gas-fired generating plants are put forward.

These considerations need to be put into the perspective of world infrastructure needs, additions to generating capacity and the investment required for both this and transmission and distribution of electricity. Table 3 shows these for 2011-2035.

2.3 FUELS FOR ELECTRICITY GENERATION TODAY

This book considers principally the question of electricity generation, which in industrialised countries accounts for about 40% of the primary energy supply. An increasing constraint on choosing the fuel for electricity generation is the carbon emissions involved.

A more fundamental consideration is the energy density of the energy source harnessed to make the electricity. Table 2 on page 9 indicates the range of energy densities from wood through coal and gas to nuclear fuel. A high-density source will generally be more efficient in use of resources to generate the electricity.

In densely populated areas of the world such as Japan and many parts of Europe and North America, coal supply is relatively remote from electricity demand. Also the high density of population and industrialisation has limited the attractiveness of coal not only from a cost but also an environmental point of view (see Chapter 8).

Therefore the desirable criteria for a fuel for base-load electricity generation in such parts of the world may be represented thus:

- The fuel should be relatively cheap, giving low-cost power.
- Unless it can be supplied from a source very close to

Figure 6. Fuel for electricity generation (2009)

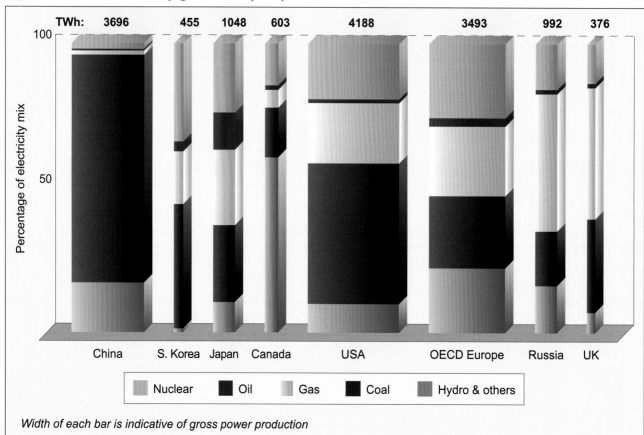

Width of each bar is indicative of gross power production

Main source: OECD/IEA Electricity Information 2011

---

[4] As a rule, individual units should not comprise more than about 15% of the system's capacity.

The Kárahnjúkar hydro-electric dam in Iceland

for this purpose. Nuclear fuel has the advantage that very little is required, and transport costs are negligible. Also, variations in the price of the fuel have very much less impact than with coal or gas (see Figure 9 on page 23).

Figure 6 shows how electricity is produced in some countries and in Europe. In all countries the demand for electric power is increasing steadily (mostly 3-4% per year). The Figure shows that coal provides a lot of the primary energy input for electricity in the USA and Europe, and slightly less in Japan and Canada. Europe, USA, Japan and South Korea have about one-fifth to one-third of their electrical power being generated from nuclear reactors.

Russia aims to increase the proportion of electricity generated by nuclear power explicitly in order to maximise gas exports to Europe. Both Russia and the UK have a high dependence on gas. The UK is keen to diminish this for energy security reasons.

the power station, it should ideally be a concentrated source of energy, which can therefore be economically transported and readily stockpiled.

- The scarcity of the resource and alternative valued applications (such as burning directly, or chemical feedstock) should be taken into account.
- Wastes should be manageable, so that they produce a minimum of pollution and environmental disturbance, including long-term global warming effect.
- Its use must be safe both in routine operation and regarding possible accident scenarios.

Of the main fuels available for base-load electricity generation, uranium often fits these criteria better overall than coal or gas, especially if the coal must be transported very far.

National energy strategies vary according to the indigenous resources of each country, the economics of importing fuels (or electricity), the amount of industrialisation, and the security of supply.

An energy-rich country such as the USA has a variety of options. However, even in parts of the USA, transporting large quantities of coal long distances adds significantly to costs.

Japan lacks indigenous energy resources and relies almost entirely on imports. Oil was once the most convenient fuel import and the country depended on it for a large proportion of its energy needs, including two-thirds of electricity generation. Coal then became increasingly used

2.4 PROVISION FOR FUTURE BASE-LOAD ELECTRICITY

In considering the future beyond a decade hence, there are a couple of practical matters which cannot be overlooked. One is the lead time. A commitment today regarding a large base-load generating plant means that plant should be commissioned in five to ten years' time. It can then be expected to have an operating life of up to 60 years. Thus today's investment decisions regarding electric plant cannot radically change the overall pattern of a country's generating system for several decades.

Combined cycle gas turbines (CCGT), which can be put into service in less than two years from date of order, and which became very popular in the 1990s, can make a significant short-term change to electricity generation but at the cost of fuel price vulnerability. If we are considering new technologies for coal-fired or nuclear base-load plant, the lead time is longer. Much of the technology in use today will inevitably be in use for several more decades – it cannot quickly be abandoned.

The other practical matter relates to size. In some things, small is appropriate and, given low labour costs, also efficient. In generating electric power however, there is economic advantage in larger plants. Where the scale is reduced, the unit costs per kilowatt inexorably increase. With conventional types of plant, large-scale installations are inevitable in urbanised and industrialised nations, where much electricity demand is concentrated in small areas of the country.

These practical matters of long lead time and large-scale installations point to the need for careful assessment of future trends in electricity use to ensure that tomorrow's supply systems will effectively cope with predicted electrical demand.

A major policy challenge is that deregulated electricity markets make the financing of any capital-intensive generating plant more difficult, even if its output is at lower cost than alternatives, and that issue has yet to be fully addressed in most countries.

Furthermore, the technology used must be matched to the task. Ultimately it is governments which must determine policies regarding the most appropriate means of generating base-load electricity for particular regions in the future. What are the options?

Conservation

One possibility may be to use less energy by practising rigorous conservation, principally through increased energy efficiency, and demand management. This approach is relevant to many applications in developed countries, and can be applied to new installations in all countries. Energy conservation in general is discussed in Section 1.6. However, such conservation has a greater effect on total energy use than on actual electricity, and an increased proportion of electricity in the overall energy mix is often a corollary of conservation.

Oil

About 5% of all electricity generation is based on burning oil, rather less than a decade earlier, and in 1973 it provided a quarter of the world's power. But oil is uniquely important as the source of very portable and energy-rich petroleum products used for transport. Both oil and gas have important uses in the petrochemical industry as feedstock for the manufacture of plastics, fertilisers and pharmaceutical products. Burning oil in a steam-cycle power plant for base-load electricity generation where other fuels are economically available is questionable. Apart from the Middle East, oil tends to be used for power generation in areas remote from natural gas resources and coalfields, in relatively small installations.

Natural gas

Gas today plays a major role in power generation (21% of world electricity in 2009). When gas prices looked like staying low and gas turbines were relatively cheap and quickly built, it was a most attractive fuel. It has the distinction of giving rise to less carbon dioxide than coal, and hence has been favoured in a short-term perspective to displace some coal for base-load power for this reason,

Goldisthal in Thuringia, Germany's largest pumped water storage plant (Photograph supplied by Vattenfall)

though the methane releases from both sources and transmission need to be factored in.

Natural gas is a superbly useful resource. It can be drawn from the Earth, easily and economically transported via large pipelines, then cheaply reticulated to small-scale points of use where it can be used as a fuel very efficiently (up to 90% at end use, allowing for flue losses). It can be liquefied for shipping to overseas customers (for example as LNG to Japan, Korea and the USA). It is also a valuable chemical feedstock for manufacturing – half the world's nitrogen fertilisers come from natural gas.

This means that large-scale use of gas for generating electricity, where less versatile alternatives are readily available, is likely to encounter acute economic constraints if gas prices rise – as in several parts of the world a few years ago, and as expected in the next decade if shale sources become less accessible. There are also ethical questions, particularly relating to intergenerational equity. In short, our grandchildren may later wish that the 1990s 'dash for gas' had been more restrained, and had left more gas for them to use in higher-value applications. Referring to domestic use rather than electricity generation, Britain's Chief Scientist said in October 2009: "Setting fire to chemicals like gas should be made a thermodynamic crime. If people want heat they should be forced to get it from (electrically-driven) heat pumps."

Methyl hydrates on the ocean floor – effectively frozen natural gas – have been suggested as a major future source of hydrocarbons, but they are far from being readily available, and concerns about methane release to the atmosphere also have to be addressed.

Coal

Of the fuels for base-load electricity generation, coal is at present the most widely-used and has been so for many years, providing 40% of the world's electricity in 2009. Modern coal-fired power stations are more efficient than in the past, and at extra cost some of the environmental effects of burning high-sulfur coals can be eliminated, even if the global warming effect due to the production of massive amounts of carbon dioxide presently cannot (see Chapter 8). However, a lot of work is being done on 'clean coal' technologies designed to reduce carbon emissions from coal burning. While these look technically feasible, the cost is likely to be very high and they have yet to be deployed commercially.

Coal from large open cut mines is fairly cheaply obtained, but the costs of transport over long distances can make it less attractive than alternatives. If large quantities of coal are mined in one locality and shipped across a continent or overseas (for example, from Australia, Canada or South America to Japan or Europe), its handling and transport imposes significant costs and involves the consumption of further energy.

Another constraint on coal is cooling the plants. Wherever possible coal-fired plants are located close to the fuel supply, and so may need fresh water for cooling. This can make a major demand on resources of fresh water.

Also, like oil and gas, coal has important uses other than as a fuel. Carbon in any grade of coal is needed in large quantities for metal smelting, for conversion to gas and liquid fuels, and for other purposes. Although known resources are large, conservation will become increasingly important.

Uranium

The only other fuel which is a present option for base-load electricity is uranium. While relatively large amounts of ore may be mined and treated, two or three 200-litre drums of uranium oxide (U_3O_8) concentrate leaving the mine contain enough energy to keep large cities supplied with power for a day, so it is very concentrated and portable.

Uranium used for nuclear power also has significant environmental advantages (see Chapter 8) which are increasingly recognised. Nuclear power is a mature technology – it is more than half a century since the first commercial reactor came on line, and almost 70 years since nuclear fission (see Chapters 3 & 10) was first controlled. In that time approximately 15,000 reactor-years of operating experience have been acquired with commercial reactors, and nearly the same from similar (but smaller) reactors in naval use.

Today, there are over 430 nuclear power reactors operable in 30 countries, including several developing nations. They provided 13.4% of the world's electricity in 2009. Many more nuclear power stations are under construction or firmly planned. The reliability, safety and economic performance of nuclear power relative to coal or oil (see also Section 2.6 and Chapter 8) has been demonstrated in many countries, especially those where at least one-fifth of their electricity is generated by nuclear power. France generates three-quarters of its electricity from nuclear power and is the world's largest electricity exporter.

Since a nuclear power station needs only a few tonnes of fresh fuel each year, location is potentially more flexible. For cooling, plants can readily be sited on the coast and use sea water, avoiding any depletion of fresh water resources.

Table 5 on page 30 gives an indication of the different kinds of nuclear power reactors currently being used for electricity generation. In the longer term, fast neutron reactors (see Section 4.6) have the potential for vastly increasing the electric power yield from uranium, though uranium resources are abundant and are not a limiting factor.

Apart from military weapons and naval propulsion, uranium has no significant uses other than for electricity generation. Making medical and industrial isotopes is an important minor use, but requires little fuel. At least 95% of the world's uranium production today goes into electricity generation (the balance to naval propulsion and isotope production – see Chapter 7).

The potential of nuclear power for electricity generation, using uranium as a fuel, is principally applicable to nations which have large blocks of electricity demand. Today's nuclear power stations tend to be built in sizes from 1000 megawatts electrical (MWe) to about 10,000 MWe, with individual units over 1000 MWe. Anything smaller currently tends to be less attractive economically, though this may change. However, there are some developing nations which have moderate-sized electricity production and distribution systems and/or the need for co-generating (for example, electricity and potable water production). These are able economically to use reactors in the 100 to 200 MWe size range where expensive oil-fired generation is the main alternative.

Nuclear fusion

As well as looking for ways to harness incident sunlight, people have for a long time dreamed of taming the process which generates that light and heat – bringing the Sun right down to Earth. That process concerned is called nuclear fusion (as distinct from fission, see Chapter 3). The favoured method for achieving controlled fusion involves joining the nuclei of deuterium and tritium atoms (heavy isotopes of hydrogen) together at very high temperatures – about 100 million degrees Celsius. No method of sustaining such temperatures under stable conditions has yet been demonstrated. However, research continues, particularly in Japan, Europe, USA and Russia, and notably in the ITER facility being built in France. Perhaps some time in the next half century heat from fusion will be harnessed to generate electricity. The technology would be best suited to large-scale base-load applications such as supplying cities and industrial regions.

Deuterium is relatively abundant in sea water, but tritium is derived either from lithium, or produced in heavy water-moderated reactors. Almost limitless energy would be available if the deuterium-deuterium reaction could be achieved, but this requires much higher temperatures than the deuterium-tritium process. Controlled fusion of ordinary hydrogen nuclei as occurs in the Sun seems unlikely ever to be achieved on Earth, as the conditions required are even more extreme. An advantage of all these reactions is that only small quantities of

radioactive wastes are expected. Disadvantages include projected high capital costs, the high radioactivity created in structural components of the plant, the cost of producing tritium, and the hazard of handling it.

2.5 RENEWABLE ENERGY SOURCES

Technology to utilise the forces of nature for doing work to supply human needs is as old as the first sailing vessel. There is a fundamental attractiveness about harnessing such forces in an age which is very conscious of the environmental effects of burning fossil fuels. Popular attention in recent years has turned to the huge sources of energy surging around us in nature – Sun, wind, and seas in particular. There was never any doubt about the magnitude of these, nor their virtuous cleanness; the challenge was always in harnessing them. Consequently a huge amount of effort, R&D and investment has gone into them.

Sun, wind, waves, rivers, tides and the heat from radioactive decay in the Earth's crust as well as biomass are all abundant and ongoing, hence the term 'renewables'. Only one, the power of falling water, has been significantly tapped for electricity so far – 16.5% of world generated power in 2009 was hydro-electric, though tidal flows and wind may perhaps one day catch up. Apart from a few countries with an abundance of it, hydro capacity is normally applied to peak-load demand, because it is so readily stopped and started. This also means that it is an ideal complement to wind power in a grid system, and Norway's capacity is used thus most effectively by interconnected Denmark, which has a high wind power capacity.

Experimental tidal current turbine, hoisted from water for maintenance

Wind is increasingly seen as a mainstream energy source. Solar energy's main human application has been in agriculture and forestry, via photosynthesis, but increasingly it is harnessed for heat. Biomass (e.g. sugar cane residue) is burned where it can be utilised, and government policies will see it increase in OECD countries. Natural geothermal power generation, tapping underground steam, is important in a few localities and hot fractured rock geothermal[5] shows promise in others. Ground source heat pump systems or engineered geothermal systems have much wider applicability, though the temperatures involved are much lower.

Olsvenne wind farm, Sweden

Tidal flow turbines would seem to have greater potential than wind to deliver power more or less continuously, but they have yet to be proven commercially.

Intermittent renewable sources

Turning to the use of less predictable intermittent renewable energy sources for electricity, there are immediate challenges in actually harnessing them and then matching this to human need. Certainly they are abundant and widespread, but apart from photovoltaic (PV) systems, the question is how to make them turn dynamos to generate the electricity. If it is heat which is harnessed, this is via a steam-driven generating system.

The fundamental problem, especially for electricity supply, is the variable and diffuse nature of solar and wind energy. This means either that there must be reliable duplicate or back-up sources of electricity, or some means of electricity storage on a large scale. Apart from pumped-storage hydro systems, no such means exist at present and nor is any in sight. For a stand-alone system the energy storage problem remains paramount. Any substantial use of solar or wind for electricity in a grid means that there must be allowance for near 100% back-up with hydro or fossil fuel capacity. This usually

gives rise to very high generating costs by present standards, but in some places it may be the shape of the future.

There are now many thousands of wind turbines operating in various parts of the world, with a total capacity of 238 GWe at the end of 2011, a doubling in three years. This has been the most rapidly-growing means of electricity generation in the last decade and provides a valuable complement to large-scale base-load power stations. Where there is an economic back-up which can be called upon at very short notice (e.g. hydro), a significant proportion of electricity can be provided from wind. The most economical and practical size of commercial wind turbines is now over 2 MWe capacity, grouped into wind farms up to 200 MWe. Some new turbines are 5 MWe. Depending on site, most turbines operate at about 25% load factor over the course of a year (European average), but some reach 33%. Wind is projected to supply 3% of world electricity in 2030, and perhaps 10% in OECD Europe. In the *New Policies* scenario in the International Energy Agency's *World Energy Outlook 2011*, 1304 GWe of new wind capacity would be added by 2035, offset by 397 GWe retired to then.

New concentrating solar thermal power stations in reliably sunny regions, using mirrors to concentrate sunlight to heat molten salt, have some potential for heat storage into the evening, thus approximating to much of the daily load demand profile. The heat is released to make steam to drive turbines.

Beyond this, intermittent renewable sources cannot be controlled to provide either continuous base-load power, or peak-load power when it is needed. In practical terms they are therefore limited to a small proportion of the capacity of an electricity grid – possibly 20%, unless coupled with nearby hydro-electric sources. They cannot directly be applied as economic substitutes for fossil fuels

---

5 Hot fractured rock (HFR) or hot dry rock technology involves pumping water down and through hot rocks, or using hot brine from deep granites some 4-5 km underground. These rocks are hot – around 250°C – because they have high levels of radioactivity and are insulated. They typically have 15-40 ppm uranium and/or thorium, but may be ten times this. The normal thermal gradient in the Earth's crust is only 25-30°C per km.

Figure 7. Fuel and waste comparison for uranium and coal

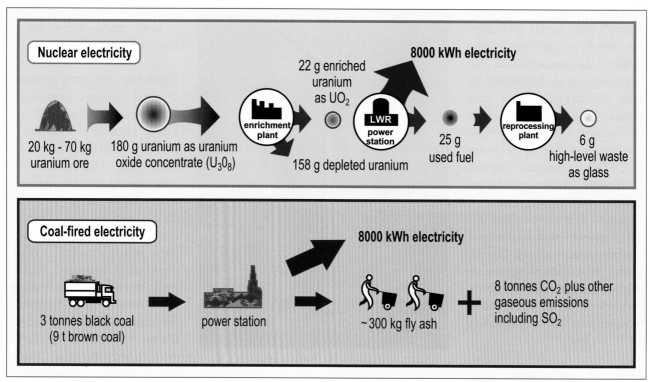

Nuclear electricity

20 kg - 70 kg uranium ore → 180 g uranium as uranium oxide concentrate (U_3O_8) → enrichment plant → 22 g enriched uranium as UO_2 → 158 g depleted uranium → LWR power station → **8000 kWh electricity** → 25 g used fuel → reprocessing plant → 6 g high-level waste as glass

Coal-fired electricity

3 tonnes black coal (9 t brown coal) → power station → **8000 kWh electricity** → ~300 kg fly ash + 8 tonnes CO_2 plus other gaseous emissions including SO_2

or nuclear power, however important they may become in particular areas with favourable conditions. Nevertheless, such technologies can and will contribute helpfully, especially where there is the political will to make consumers subsidise them, even though they are unsuitable for carrying the main burden of supply.

A great deal of investment is going into both solar photovoltaic (PV) and concentrating solar thermal power (CSP) today, though even more than wind this relies on subsidies or higher prices from consumers. In the *New Policies* scenario of the *World Energy Outlook 2011*, 553 GWe of new solar PV and 81 GWe of CSP capacity would be added by 2035. Solar PV capacity at the end of 2011 was 67 GWe.

If there were some way that large amounts of electricity from intermittent producers such as solar and wind could be stored efficiently, the contribution of these technologies to supplying actual electricity demand would be much greater. Making hydrogen by electrolysis which is then used in fuel cells is a possibility, but appears remote practically.

Already in some places pumped storage is used to even out the daily generating load by pumping water to a high storage dam during off-peak hours and weekends, using the excess base-load capacity

from coal or nuclear sources. During peak hours this water can be used for hydro-electric generation feeding the grid. Relatively few places have scope for pumped storage dams close to where the power is needed, and overall efficiency is low (around 65%). Means of storing large amounts of electricity as such in giant batteries or by other means have not been developed.

Turbine hall at Olkiluoto nuclear power plant, Finland

Environmental aspects of renewables

Renewable energy sources have a completely different set of environmental costs and benefits to fossil fuel or even nuclear generating capacity.

On the positive side they emit no carbon dioxide or other air pollutants (beyond some decay products such as methane from new hydro-electric reservoirs), but because they are harnessing relatively low-intensity energy, their 'footprint' – the area taken up by them – is necessarily much larger. Whether large areas near cities dedicated to solar collectors will be acceptable (if such proposals are ever made) remains to be seen. Beyond utilising roofs, 1000 MWe of solar capacity would require at least 20 square kilometres of collectors, shading a lot of country to the extent that agricultural productivity would be minimal.

In Europe, wind turbines have not endeared themselves to neighbours on aesthetic, noise or nature conservation grounds, and this has arrested their onshore deployment particularly in the UK. At the same time, European non-fossil fuel obligations have led the establishment of major offshore wind farms and the prospect of more.

However, much environmental impact can be reduced. Fixed solar collectors can double as noise barriers along highways, roof-tops are available already, and there are places where wind turbines would not obtrude unduly.

See also: WNA information paper on *Renewable Energy and Electricity*.

2.6 COAL AND URANIUM COMPARED

The only major fuel options for large-scale energy conversion to base-load electricity over the next several decades are coal and uranium.

Gas is an option in some places in the short term, and while its price remains low it is very attractive compared with high capital cost alternatives. However, its great value as a direct fuel and the likelihood of significant price volatility put the spotlight back onto coal and uranium.

Coal-fired heat and power plant located in Pruszków, Poland (Photograph supplied by Vattenfall)

Choices between these alternatives will probably continue to depend principally on the final cost of electric power (including costs associated with carbon emissions and other environmental costs), which varies significantly from site to site, and in any case needs to be projected forward at least a couple of decades. Financing the immediate capital expenditure involved will also be a factor.

Some general comparisons between coal and uranium as the principal fuels for base-load electricity generation are discussed in this section. Other comparisons which are principally environmental or related to health, *i.e.* external costs, are discussed in more detail in Chapter 8.

Different quantities of materials are involved with energy conversion to electricity, starting with coal and uranium. In either case the amount of electricity considered is 8000 kWh, a conservative estimate of the amount required by one person in a developed country for one year.[6]

Using uranium as the fuel

For 8000 kWh, between 20 kg and 70 kg of uranium ore from a typical mine is needed to produce a handful (210 grams) of uranium oxide concentrate. The uranium in this concentrate, is referred to as 'natural uranium' and contains about 0.7% U-235, the fissile isotope of uranium. For most nuclear reactors the natural uranium is enriched in its U-235 isotope to yield about 25 grams of enriched uranium fuel (say 4% U-235, see Section 5.2).

Used fuel from nuclear reactors contains a significant quantity of fissile material and in some countries it is

---

[6] The average consumption in OECD countries is 7640 kWh/yr (*World Energy Outlook 2008*). Canadian consumption is 15,300 kWh/person/year, European consumption is 5600 kWh/person/year and in the USA it is 12,420 kWh/person/year (OECD/IEA *Electricity Information 2008*).

reprocessed to recover this for recycle. When used fuel is reprocessed, most of it is recycled, and the tiny amount of separated waste is incorporated into about five grams of pyrex glass – about the size of a small coin. However, it is highly radioactive. Other wastes are also produced, but they are of much less significance (see Section 6.1).

Using coal as the fuel

About three tonnes of high quality black coal (or 3.5 t of average black coal or up to 9 tonnes of brown coal) can be fed into a power station to generate the same amount of electricity – 8000 kWh. This leaves a certain amount of ash, varying from a couple of barrow loads to half a tonne, depending on the particular coal used. Eight tonnes of carbon dioxide (CO_2), which at atmospheric temperature and pressure would fill three full-sized Olympic pools (50m x 15m x 2m), is produced. For the foreseeable future, this is discharged into the atmosphere.

Depending on the coal, some sulfur dioxide (SO_2) is also produced. A common type of US coal might contain 2-3% sulfur, in which case possibly a hundred kilograms of sulfur dioxide would require costly removal, or would add to the acid rain problems well known in the northern hemisphere. (Australian and Canadian coal generally contains less than 1% sulfur.) The environmental effects of these gaseous by-products of coal-fired electricity generation are considered in more detail in Sections 8.1 and 8.2, and the costs of SO_2 removal are significant.

Years ago, most coal-fired power plants emitted more radioactivity than any nuclear plants of similar size! This was due to trace quantities of radioactive materials (e.g. often about 5 ppm uranium plus thorium in Australia & Canada) in the coal. With modern equipment this radioactivity is mostly retained with the ash and is buried with it.

2.7 ENERGY INPUTS TO GENERATE ELECTRICITY

Any electricity generation requires some energy inputs in mining, concentrating and transporting the fuel, manufacturing and constructing the plant, and dealing with the wastes. Energy use in mining and transport is closely related to quantities involved, and any comparison therefore favours uranium. On the other hand, the capital-intensive nature of the nuclear fuel cycle is reflected in the plant, and the greater energy inputs to it.

The main energy input to the nuclear fuel cycle for reactors requiring enriched fuel may be in enriching uranium (see Section 5.2). The following figures consider

The Loy Yang lignite (brown coal) mine in Australia covers 6 square kilometres after 25 years

Figure 8. US electricity production costs

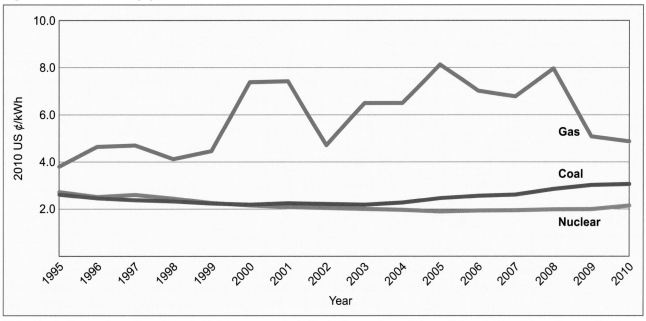

Note: Production costs = Operations & Maintenance + Fuel. Production costs exclude capital cost since this varies greatly among utilities and states.
Source: Nuclear Energy Institute

Figure 9. Components of electricity costs – projected electricity costs for Finland

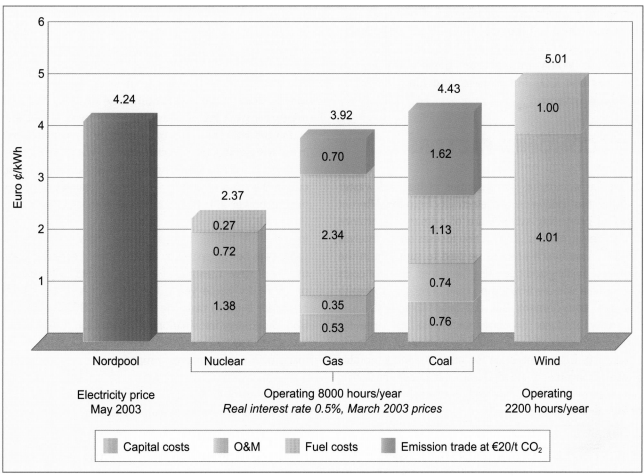

Source: R. Tarjanne & K. Luostarinen, 2003, Lappeenranta University of Technology

a 1000 MWe reactor run at 90% and therefore generating almost 8 billion kWh/yr. Conservatively, this would require about 170 tonnes of natural uranium each year, which might be enriched to produce 17.5 tonnes of uranium oxide fuel containing 16 tU at 4% U-235. On the way to becoming fuel, this natural uranium, after conversion to uranium hexafluoride (UF_6), would need 7 GWh of electricity to enrich it in a modern centrifuge plant[7].

Then there is fuel fabrication as well as construction and operation of the reactor to include. The total energy inputs to the nuclear fuel cycle over a nuclear plant's full lifetime represent about 1.7% of the energy output[8]. The energy payback time for a nuclear plant is about four months at full output.

Mining, at Ranger (open cut) or Beverley (in situ leaching), uses energy equivalent to 0.06% and 0.05% respectively of the mine's output if that uranium is used in a light water reactor. If extremely low-grade uranium ore (0.01% U) is assumed, the life-cycle energy input rises to nearly 4%.

Life-cycle figures for coal range from 3.5 to 7% of the energy output required for inputs. In mass terms, the fuel inputs provide a stark contrast. Compared with uranium, about 20,000 times as much coal is needed.

See also: WNA information papers on *Energy Balances and CO$_2$ Implications* and *Energy Analysis of Power Systems*.

2.8 ECONOMIC FACTORS

As well as comparing the quantities of fuel and wastes involved, the relative costs of electricity from the different types of generating systems are of primary importance in considering options. This section focuses on the internal costs – those which need to be paid to suppliers in the course of building and operating the plants. External costs are those which are actually incurred in relation to health and the environment but not paid directly by the electricity producer or consumer. These are large for fossil fuels, especially coal, and are considered further in Chapter 8. Equivalent costs for nuclear energy, notably waste management and disposal, and decommissioning old reactors, are internalised and paid for by the consumers of their electricity.

A nuclear power station costs a lot more than a gas-fired station and somewhat more than a coal-fired station to build. But the nuclear fuel, including enrichment if needed, costs much less than oil, gas or coal. Hence the overall cost for energy conversion to electricity can come out much the same for nuclear as for coal-fired plants. Table 4 quotes some comparisons of reported electricity costs compiled by the OECD, and Figure 8 shows the actual costs (excluding capital) over more than a decade in the USA, while Figure 9 shows the components of electricity cost for different means of generating it.

There are a number of US nuclear plants where capital costs overran badly during construction in the 1970s-80s and hence where any normal calculation of generating cost shows it to be very high. However, the criterion for running them is the cost of actual operation (O&M plus fuel – see Figure 8). On this basis they compare favourably with coal and are cheaper than gas. Nearly 20 of these older US reactors changed hands over 1998-2005 and the escalating prices indicated the favourable economics involved. Regarding investment in new capacity, the capital costs are a major factor, and these are included in Table 4 and Figure 9.

In an earlier version of Table 4, OECD figures for plants starting operation in 2000 showed the importance of having coal near its point of use and low in sulfur (flue gas desulfurisation is costly both in capital and energy terms). Costs in the northeastern USA distinctly favoured nuclear; costs in the midwest marginally favoured nuclear; and in the west, coal was often cheaper. This is reflected in today's distribution of nuclear plants. Having the location of electricity demand a long way from sources of cheap coal is the main reason for the steadily increasing use of nuclear power in many countries as compared with coal.

In the USA today, low gas prices show new nuclear capacity as being apparently uneconomic. However, given that the new tranche of nuclear plants will not be operating for some years, the question is what the gas prices are likely to be in the 2020s and 2030s. Projections from the Energy Information Administration (EIA) in 2012 show gas prices in the 2020s rising to levels which will make new nuclear plants competitive.

Generally plant choice is likely to depend on a country's international economic situation. Nuclear power is very

---

[7] At a tails assay of 0.25% U-235 in the enrichment plant, 5.9 SWU per kg of 4% enriched product is required, @ 50 kWh/SWU for the modern centrifuge plant (or up to 2400 kWh/SWU for the older gaseous diffusion plant). The 175 tonnes of natural uranium would leave the mine as 207 t U$_3$O$_8$.

[8] With older diffusion enrichment the figure could be up to 5%.

Table 4. Actual costs of electricity (US ¢/kWh)

Technology	Region or country	At 10% discount rate	At 5% discount rate
Nuclear	OECD Europe	8.3 - 13.7	5.0 - 8.2
	China	4.4 - 5.5	3.0 - 3.6
Black coal with CCS	OECD Europe	11.0	8.5
Brown coal with CCS	OECD Europe	9.5 - 14.3	6.8 - 9.3
CCGT with CCS	OECD Europe	11.8	9.8
Large hydro-electric	OECD Europe	14.0 - 45.9	7.4 - 23.1
	China: Three Gorges Dam	5.2	2.9
	China: other	2.3 - 3.3	1.2 - 1.7
Onshore wind	OECD Europe	12.2 - 23.0	9.0 - 14.6
	China	7.2 - 12.6	5.1 - 8.9
Offshore wind	OECD Europe	18.7 - 26.1	13.8 - 18.8
Solar photovoltaic	OECD Europe	38.8 - 61.6	28.7 - 41.0
	China	18.7 - 28.3	12.3 - 18.6

Source: OECD/IEA-NEA, 2010, Table 3.7 of Projected Costs of Generating Electricity
This shows the levelised cost, which is the average cost of producing electricity including capital, finance, owner's costs on site, fuel and operation over a plant's lifetime.

capital-intensive (see Figure 9), while fuel costs are relatively much more significant for systems based on fossil fuels. Therefore if a country such as Japan or France has to choose between importing large quantities of fuel or spending a lot of capital at home, the actual upfront costs may be less important than wider economic considerations. Development of nuclear power, for instance, provides work for local industries which build the plant and also minimises long-term commitments to buying fuels abroad. Overseas purchases of fossil fuels for new electricity plant would give rise to decades of economic and supply vulnerability.

Uranium has the advantage of being a highly concentrated source of energy which is therefore easily and cheaply transportable. The quantities needed are very much less than for coal[9] (see Section 2.6). In addition, the fuel's contribution to the overall cost of electricity produced is relatively small, which means that even a large fuel price escalation will have relatively little effect.

However as the long-term global environmental consequences of consuming fossil fuels, especially coal, create additional concern, the environmental advantages of nuclear power are also receiving more attention (see Chapter 8) and will increasingly be reflected in the overall economics if costs are imposed on carbon emissions.

Assigning carbon values, or imposing carbon costs, on fossil fuel electricity generation changes the economic situation relative to nuclear energy. For instance, carbon values of $10 per tonne of CO_2 for typical coal will increase the electricity cost from that source by one cent per kilowatt hour while leaving nuclear electricity costs unaffected.

Although coal and uranium broadly compete for base-load electricity generation, most developed nations fortunate enough to have the option see a role for both.

As a general rule, countries without cheap coal or plentiful gas tend to favour nuclear power as the lower cost option. In a few countries (such as Australia, where coal reserves and production potential far outweigh domestic needs) the use of coal for electricity generation is favoured over nuclear. However, in a world perspective, the need for both is evident, and as electricity demand increases along with concerns regarding possible global warming, not to mention energy security, a corresponding preference for nuclear power to generate base-load electricity seems inevitable in many countries.

See also: WNA information paper on *The Economics of Nuclear Power* and *World Energy Needs and Nuclear Power*.

---

[9] One kilogram of natural uranium yields about 20,000 times as much energy as the same amount of coal (see Table 2 in Chapter 1).

3. Nuclear power and its fuels

3.1 MASS TO ENERGY IN THE REACTOR CORE

Until relatively recently people must have thought they were converting mass to energy when they burned wood to cook meals and to keep warm, but any student today would be aware that this was not the case. One form of carbon compound (the solid wood) was simply being converted to another (a colourless gas), which blew away. The hydrogen associated with the original compound also dispersed as water vapour. No measurable mass was lost, although energy was released. However, during the 20th Century, as our understanding of nuclear physics developed, it was recognised by Albert Einstein that mass could in fact be turned into energy. This is what happens in a nuclear reactor, using atoms of particular metals such as uranium.

Uranium is 1.7 times more dense than lead, and is composed of atoms which have in their nucleus 92 protons (positively-charged) and about 140 neutrons (uncharged). One of the types of uranium atoms, or uranium 'isotopes' as they are called, has 143 neutrons. This uranium-235 (U-235) isotope is remarkable because when its nucleus is hit by a slow neutron (also known as a 'thermal' neutron) the atom can split in two and release a lot of energy as heat. This is called nuclear 'fission', and U-235 is thus a 'fissile' isotope. In Einstein's terms, some mass is converted to energy. At the same time, several fast neutrons are emitted from the split nucleus. If these are slowed by a moderator such as graphite or water they can cause other U-235 atoms to split, thus giving rise to a chain reaction (see also Section 3.7).

The most common isotope of natural uranium, U-238, is not itself fissile in conventional reactors but it can become fissile plutonium-239 through neutron capture. It is thus 'fertile'. Pu-239 is fissile and behaves similarly to U-235. About one-third of the energy from a commercial nuclear

Figure 10. Fission in conventional and fast neutron reactors

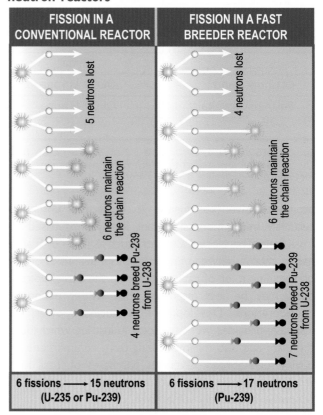

Contrast between conventional ('thermal') reactor and fast neutron reactor showing how typically more neutrons are produced in the fast reactor (17 instead of 15 from 6 fissions), thus enabling the system to breed more fissile material than is consumed if desired. In this example, four neutrons are available for breeding Pu-239 in the conventional reactor, but seven are available in the fast reactor. The exact numbers involved will depend on design and operation.

reactor comes from fission of the Pu-239 produced in the reactor, and about two-thirds from the U-235.

The reactor core is loaded with uranium oxide fuel. In conventional reactors this is enriched to 4-5% U-235 (see also Section 5.2)[1]. It is typically in the form of ceramic pellets of uranium dioxide (UO_2) with melting point of about 2800°C. These are assembled inside long zirconium

[1] In CANDU reactors, natural uranium – 0.7% U-235 – is used.

NUCLEAR POWER AND ITS FUELS

alloy or stainless steel tubes and surrounded by coolant and moderator.

The moderator slows down the fast neutrons from the nuclear fission reaction so that they are more likely to cause further fission in U-235 atoms. If normal (light) water is used as a moderator, the fuel must be enriched in the level of U-235 because of the tendency of water to also capture neutrons. If graphite or heavy water moderator is used, natural uranium (with just 0.7% U-235) will work as fuel.

Each U-235 fission typically releases more than a million times as much energy as from chemical reactions such as combustion. Commercial nuclear power generation involves containing and controlling the fission reactions so that the heat can be used to make steam, which in turn generates electricity. Removing the heat reliably is vital. Not only does the heat from fission need to be removed as the reactor operates, but also afterwards. For some time after shutting down, the decay heat from fission product radionuclides is significant and must also be removed from the core, as was demonstrated in the Fukushima accident.

Fuel assembly for nuclear reactor (Westinghouse)

As fission takes place in the core, the fuel changes. Its fissile content diminishes as burn-up proceeds, and new elements – both fission products and transuranic elements including plutonium – build up. While Pu-239 is a welcome supplement to U-235, many of these new elements are neutron absorbers, so progressively make the fuel less efficient. Compensation for this is provided by progressively withdrawing control rods or reducing boron[2] levels in the coolant, but at some stage – after about four years – the used fuel needs to be replaced. Typically one-third of the fuel is replaced in each refuelling outage.

The nuclear fuel cycle – the sequence of what is done to the fuel before it is used in the reactor and what happens to it afterwards – is described in Section 5.2. A fuller account of the physics involved is in Section 3.7.

3.2 NUCLEAR POWER REACTORS – BASIC DESIGN AND FUNCTION

In the middle of the last century an extraordinary variety of experimental nuclear reactors were built and operated, with every conceivable type of fuel, moderator and coolant. Gas-cooled graphite-moderated reactors were popular initially, but very soon the focus shifted to designs moderated by light water and using enriched uranium. These also predominated in naval use.

In fact, reactors derived from designs originally developed for propelling submarines and large naval ships generate about 85% of the world's nuclear electricity. The main design is the pressurised water reactor (PWR) which has water at over 300°C under pressure in its primary cooling/heat transfer circuit, and generates steam in a secondary circuit in a heat exchanger (steam generator) (see Figure 11). The less popular boiling water reactor (BWR) makes steam in the primary circuit above the reactor core. Both types use water as both coolant and moderator, to slow neutrons.

In the reactor core, uranium undergoes fission so that a lot of heat is released. The control rods regulate the rate of the reaction, and therefore the heat yield, by absorbing some of the moving neutrons. The core is immersed in water and is enclosed in a very thick steel pressure vessel.

---

[2] Boron is a neutron absorber. See Panel on next page.

Figure 11. Pressurised water reactor (PWR)

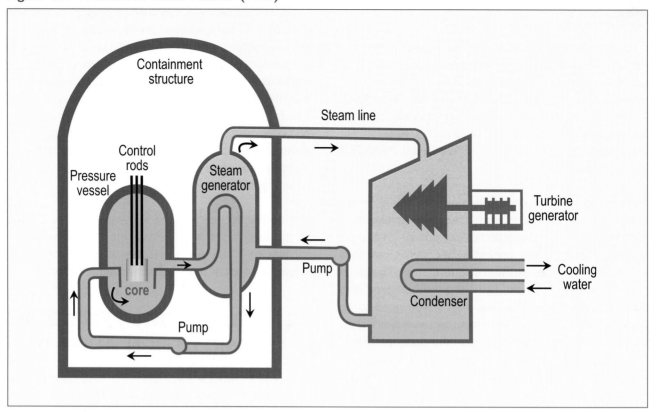

Components common to most types of nuclear reactor

Fuel. Uranium is the basic fuel. Usually pellets of uranium oxide (UO_2) are arranged in tubes to form fuel rods. The rods are arranged into fuel assemblies in the reactor core.

Moderator. Material in the core which slows down the neutrons released from fission so that they cause more fission. It is usually water, but may be heavy water or graphite.

Control rods. These are made with neutron-absorbing material such as cadmium, hafnium or boron, and are inserted or withdrawn from the core to control the rate of reaction, or to halt it. In some PWR reactors, special control rods are used to enable the core to sustain a low level of power efficiently. (Secondary control systems involve other neutron absorbers, usually boron in the coolant – its concentration can be adjusted over time as the fuel burns up.)

Coolant. A fluid circulating through the core so as to transfer the heat from it. In light water reactors the water moderator functions also as primary coolant. Except in BWRs, there is a secondary coolant circuit where the water becomes steam. (The steam is used to drive a turbine to make electricity.)

Pressure vessel or pressure tubes. Usually a robust steel vessel containing the reactor core and moderator/coolant, but it may be a series of tubes holding the fuel and conveying the coolant through the surrounding moderator.

Steam generator (not in BWR). Part of the cooling system where the primary coolant bringing heat from the reactor is used to make steam in a secondary circuit for the turbine. Essentially a heat exchanger like a motor car radiator. Reactors may have up to four 'loops', each with a steam generator.

Containment. The structure around the reactor and associated steam generators which is designed to protect it from outside intrusion and to protect those outside from the effects of radiation in case of any serious malfunction inside. It is typically a metre-thick concrete and steel structure.

Reactor pressure vessel being installed at Ling Ao nuclear power plant, China. Three steam generators surround it. (CGNPC)

to the Fukushima accident and Germany's decision to close eight older reactors in 2011 for political rather than safety reasons, countered what would otherwise have been a positive year in added nuclear capacity with 4000 MWe coming on line.

Secondly, increased nuclear capacity in some countries is resulting from the uprating of existing plants. Power reactors in the USA, Belgium, Sweden, Spain, Switzerland and Germany, for example, have had their generating capacity increased. In 2011 there were about nine uprates in four countries, adding 620 MWe.

Thirdly, capacity or load factors are improving everywhere, so that more kilowatt hours come from the installed capacity. More than two-thirds of the nuclear plants in the last few years have had load factors over 75%, up from 67% average load factor in 1992. The average for the two main light water types is now over 80% and many countries average over 80% load factor. US nuclear power plant performance, at over 90%, has moved into the top bracket. Recently the annual improvement in US reactor performance was equivalent to putting significant new power station capacity on line each year. To put it another way, the US increase from 65% load factor in 1980s to almost 90% today is equivalent to adding 23,000 MWe capacity.

Fourthly, plant lives are being extended. Most nuclear power plants originally had a nominal design lifetime of 30 to 40 years, but engineering assessments have established that many plants can operate longer. Extending reactor operating life by replacing major components is often an attractive and cost-effective option for utilities. In the USA and Japan, most reactors had confirmed life-spans of 40 years, but many – now 70% of those in the USA – have now been cleared to operate for 60 years. When the oldest commercial nuclear power stations in the world, Calder Hall and Chapelcross in the UK, were built in the 1950s, it was assumed that they would have a useful lifetime of 20 years. They were later assessed as being safe to operate for 50 years, though in fact they closed earlier for economic reasons.

New reactor start-ups will exceed the decommissioning of old reactors for several years at least, though most of the new reactors will be in the Asian region.

All this occurs in a big concrete and steel containment structure. The steam is fed to a turbine generator, much the same as those installed in coal-fired power stations. The uranium-fuelled core of a nuclear power reactor simply takes the place of a boiler or furnace burning coal (or other fossil fuel) to generate the steam.

Burnable poisons are often used (especially in BWR) in fuel or coolant to even out the performance of the reactor over time, from fresh fuel being loaded to refuelling. These are neutron absorbers which decay under neutron exposure, compensating for the progressive build-up of neutron absorbers in the fuel as it is burned. The best known is gadolinium, which is a vital ingredient of fuel in naval reactors where installing fresh fuel is very inconvenient, so reactors are designed to run more than a decade between refuellings.

Increasing nuclear capacity

In 2011 nuclear electricity generation was 2518 billion kilowatt hours, more than all electricity generated worldwide five decades ago, and an increase of 20% over the previous 15 years. The reasons for the overall growth are several.

First, and most obviously, capacity is slowly increasing as new reactors come on line, as suggested by Table 5. In May 2012 there were 433 nuclear power reactors with a capacity of some 372,000 MWe operating in 30 countries, with 63 power reactors (62 GWe) under construction in 14 countries and over 160 more units firmly planned. New reactor start-ups are partly offset by the closure of old plants, most of them smaller than those starting up. Having four reactors written off due

Table 5. Nuclear power's role in electricity production

	Nuclear Generation 2011		Operable at June 2012		Construction at June 2012		Planned at June 2012		Proposed at June 2012	
	TWh	% elec.	No.	MWe net	No.	MWe gross	No.	MWe gross	No.	MWe gross
Argentina	5.9	5	2	935	1	745	2	773	1	740
Armenia	2.4	33.2	1	376	0	0	1	1060	0	0
Bangladesh	0	0	0	0	0	0	2	2000	0	0
Belarus	0	0	0	0	0	0	2	2000	2	2000
Belgium	45.9	54	7	5943	0	0	0	0	0	0
Brazil	14.8	3.2	2	1901	1	1405	0	0	4	4000
Bulgaria	15.3	32.6	2	1906	0	0	1	950	0	0
Canada	88.3	15.3	17	12044	3	2190	3	3300	3	3800
Chile	0	0	0	0	0	0	0	0	4	4400
Chinese mainland	82.6	1.8	15	11881	26	27640	51	57480	120	123000
Taiwan	40.4	19	6	4927	2	2700	0	0	1	1350
Czech Republic	26.7	33	6	3764	0	0	2	2400	1	1200
Egypt	0	0	0	0	0	0	1	1000	1	1000
Finland	22.3	31.6	4	2741	1	1700	0	0	2	3000
France	423.5	77.7	58	63130	1	1720	1	1720	1	1100
Germany	102.3	17.8	9	12003	0	0	0	0	0	0
Hungary	14.7	43.2	4	1880	0	0	0	0	2	2200
India	28.9	3.7	20	4385	7	5300	16	14300	40	49000
Indonesia	0	0	0	0	0	0	2	2000	4	4000
Iran	0	0	1	915	0	0	2	2000	1	300
Israel	0	0	0	0	0	0	0	0	1	1200
Italy	0	0	0	0	0	0	0	0	10	17000
Japan	156.2	18.1	50	44396	3	3036	10	13772	5	6760
Jordan	0	0	0	0	0	0	1	1000	0	0
Kazakhstan	0	0	0	0	0	0	2	600	2	600
Korea DPR (N)	0	0	0	0	0	0	0	0	1	950
Korea RO (S)	147.8	34.6	23	20787	3	3800	6	8400	0	0
Lithuania	0	0	0	0	0	0	1	1350	0	0
Malaysia	0	0	0	0	0	0	0	0	2	2000
Mexico	9.3	3.6	2	1600	0	0	0	0	2	2000
Netherlands	3.9	3.6	1	485	0	0	0	0	1	1000
Pakistan	3.8	3.8	3	725	2	680	0	0	2	2000
Poland	0	0	0	0	0	0	6	6000	0	0
Romania	10.8	19	2	1310	0	0	2	1310	1	655
Russia	162	17.6	33	24164	10	9160	17	20000	24	24000
Saudi Arabia	0	0	0	0	0	0	0	0	16	17000
Slovakia	14.3	54	4	1816	2	880	0	0	1	1200
Slovenia	5.9	41.7	1	696	0	0	0	0	1	1000
South Africa	12.9	5.2	2	1800	0	0	0	0	6	9600
Spain	55.1	19.5	8	7448	0	0	0	0	0	0
Sweden	58.1	39.6	10	9399	0	0	0	0	0	0
Switzerland	25.7	40.8	5	3252	0	0	0	0	3	4000
Thailand	0	0	0	0	0	0	0	0	5	5000
Turkey	0	0	0	0	0	0	4	4800	4	5600
Ukraine	84.9	47.2	15	13168	0	0	2	1900	11	12000
UAE	0	0	0	0	0	0	4	5600	10	14400
UK	62.7	17.8	16	10038	0	0	4	6680	9	12000
USA	790.4	19.2	104	101607	1	1218	11	13260	19	25500
Vietnam	0	0	0	0	0	0	4	4000	6	6700
WORLD	**2,518**	**c 13.5**	**433**	**371,422**	**63**	**62,174**	**160**	**179,655**	**329**	**373,255**

Operable = Connected to the grid; Construction = first concrete for reactor poured, or major refurbishment under way; Planned = approvals, funding or major commitment in place, mostly expected in operation within 8-10 years, or construction well advanced but suspended indefinitely; Proposed = specific program or site proposals, expected operation mostly within 15 years. New plants coming on line are balanced by old plants being retired. Over 1996-2009, 43 reactors were retired as 49 started operation. There are no firm projections for retirements over the period covered by this Table, but WNA estimates that at least 60 of those now operating will close by 2030, most being small plants. The 2011 Market Report reference case has 156 reactors closing by 2030, and 298 new ones coming on line.

TWh = Terawatt-hours (billion kilowatt-hours), MWe = Megawatt (electrical as distinct from thermal), kWh = kilowatt-hour.

Sources: WNA to 1/6/12; IAEA for nuclear electricity production and percentage of electricity (% elec.).

Control room at Olkiluoto nuclear power plant, Finland (TVO)

Table 6. Uranium concentrations in nature

High-grade orebody	20,000 ppm U (2% U)
Low-grade orebody	1,000 ppm U (0.1% U)
Lowest grade sometimes mineable	100 ppm U
Granite	3-5 ppm U
Sedimentary rock	2-3 ppm U
Average in the Earth's continental crust	2.8 ppm U
Sea water	0.003 ppm U

ppm = parts per million

3.3 URANIUM AVAILABILITY

Uranium is a metal approximately as common as tin or zinc, and it is a constituent of most rocks and even of the sea. Some typical concentrations are shown in Table 6.

An orebody is, by definition, an occurrence of mineralisation from which the metal is economically recoverable. It is therefore relative to both costs of extraction and market prices. At present, neither the oceans nor any granites are orebodies, but conceivably either could become so if uranium prices were to rise sufficiently.

Measured resources of uranium, the amount known to be economically recoverable from orebodies, are thus also relative to costs and prices. They are also dependent on the intensity of exploration effort, which has been low when uranium prices were low, e.g. from the early 1980s to 2005. Published resource figures are basically a statement about what is known, rather than what is actually there in the Earth's crust.

Changes in costs or prices, or further exploration, may alter known resource figures markedly. Thus, any predictions of the future availability of any mineral, including uranium, which are based on current cost and price data and current geological knowledge, are likely to be extremely conservative.

From time to time, concerns are raised that the known resources might be insufficient when judged as a multiple of present rate of use. But this is the 'Limits to Growth' fallacy, a major intellectual blunder recycled from the 1970s, which takes no account of the very limited nature of the knowledge we have at any time of what is actually in the Earth's crust. Our knowledge of geology is such that we can be confident that identified resources of metal minerals are a small fraction of what is there.

With those major qualifications, Table 7 gives some idea of our present understanding of uranium resources.

Presently-known resources of uranium are enough to last more than half a century, considering only the lower cost category (recoverable to US$ 80/kg), and with it being used only in conventional reactors. This represents a higher level of assured resources than is normal for most minerals.

Table 7. Known recoverable resources of uranium

	Tonnes U	% of world
Australia	1,673,000	31
Kazakhstan	651,000	12
Canada	485,000	9
Russian Fed.	480,000	9
South Africa	295,000	5.5
Namibia	284,000	5
Brazil	279,000	5
Niger	272,000	5
USA	207,000	4
China	171,000	3
Jordan	112,000	2
Uzbekistan	111,000	2
Ukraine	105,000	2
India	80,000	1.5
Mongolia	49,000	1
Other	150,000	3
World total	**5,404,000**	**100**

Source: Reasonably Assured Resources plus Inferred Resources, recoverable to US$ 130/kg U, 1/1/09, from OECD NEA & IAEA, Uranium 2009: Resources, Production and Demand ('Red Book').

Further exploration and higher prices will certainly, on the basis of present geological knowledge, yield further resources as present ones are used up. A doubling of price from present contract levels could be expected to create about a ten-fold increase in measured resources, over time.

It is clear from Figure 12 that known uranium resources have increased three-fold since 1975, in line with expenditure on uranium exploration. (The decrease in the decade 1983-93 is due to some countries tightening their criteria for reporting. If this were carried back two decades, the lines would fit even more closely.) Increased exploration expenditure in the future is likely to result in a corresponding increase in known resources.

Widespread use of the fast neutron reactor (see Section 4.6) could hugely increase the utilisation of uranium. This type of reactor can be started up on plutonium derived from conventional reactors and operated in closed circuit with its reprocessing plant. Such a reactor, supplied with natural uranium as fertile material, very quickly reaches the stage where each tonne of ore yields 60 times more energy than in a conventional reactor.

See also: WNA information paper on *Supply of Uranium*.

Reactor fuel requirements

The world's power reactors, with combined capacity of 372 GWe, require some 80,000 tonnes of uranium oxide concentrate from mines (or stockpiles) each year. While this capacity is being run more productively, with higher capacity factors and reactor power levels, the uranium fuel requirement is increasing but not necessarily at the same rate. The factors increasing fuel demand are offset by a trend for higher burn-up of fuel and other efficiencies, so demand is steady. (Over the 18 years to 1993, the electricity generated by nuclear power increased 5.5 times while uranium used increased only just over three-fold.) However, it is likely that the annual uranium demand will grow in line with net increase in nuclear generating capacity as new plants come on line.

Fuel burn-up is measured in megawatt days per tonne U (MWd/t), and many operators are increasing the initial enrichment of their fuel (e.g. from 3.3% to 4.0% U-235) and then burning it longer or harder to leave only 0.5% U-235 in the fuel. This might mean that typical burn-up is increased from 33,000 MWd/t to 50,000 MWd/t.

Reprocessing of used fuel from conventional light water reactors to recycle the left-over uranium and plutonium

'Yellowcake', the penultimate uranium compound in U_3O_8 production

Figure 12. Known uranium resources and exploration expenditure

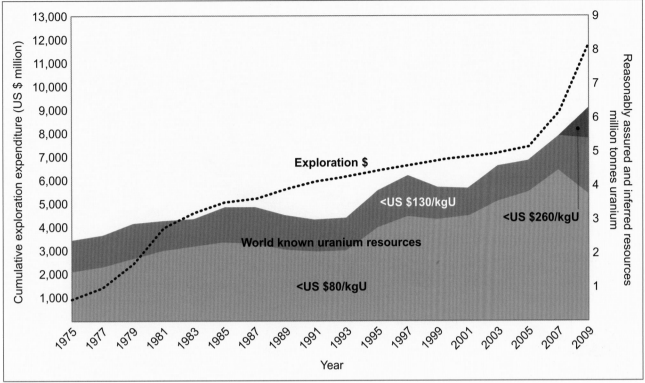

Investment in geological exploration results in increased known uranium resources.
Source: OECD NEA & IAEA, Uranium: Resources, Production and Demand ('Red Book') series from 1975.

(see Section 6.2) also utilises present resources more efficiently, by a factor of up to 1.3 overall. At present the plutonium arising from reprocessing is used in fresh mixed oxide (MOX) fuel, with depleted uranium from enrichment plants.

The net result from all this is a small reduction in the amount of uranium required ex-mine to fuel each kilowatt hour produced.

Using fission in uranium exploration

As well as geochemical methods, most uranium exploration uses the measurement of gamma rays to indicate the presence of uranium mineralisation. Gamma rays are emitted when many radioactive nuclei decay. However, in a uranium orebody this gamma emission comes from decay products, not uranium itself (see Appendix 2). Where the uranium has been leached from the original orebody leaving behind its decay products while it is deposited elsewhere, in buried river channels for instance, gamma measurements do not give a good indication of uranium concentrations. The best indication is from causing a little fission in the uranium itself.

A portable prompt fission neutron (PFN) logging tool lowered down a borehole employs a neutron source and a neutron detector. The neutron source irradiates the uranium deposit, and prompt or delayed neutrons resulting from fission of any uranium present in the formation are detected and recorded. This is the only reliable geophysical measurement of some uranium deposits.

3.4 NUCLEAR WEAPONS AS A SOURCE OF FUEL

Since 1987, the USA and countries of the former USSR have signed a series of disarmament treaties to reduce the nuclear arsenals of the signatory countries by approximately 80%. The weapons concerned contain a great deal of uranium enriched to over 90% U-235 (i.e. about 25 times the proportion in most reactor fuel). Some weapons have plutonium-239, which can be used in diluted form in either conventional or fast breeder reactors.

An important source of nuclear fuel is the world's nuclear weapons stockpiles.

Figure 13. World uranium production and demand

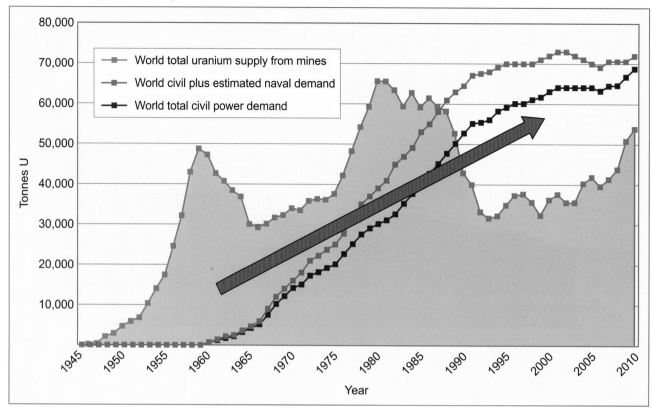

Early uranium production from mines was largely for nuclear weapons. This graph shows escalating demand for civil nuclear power production, and adds an estimate of demand for naval nuclear propulsion systems. It also shows (arrow) how some material from military stockpiles has augmented the mine supply since mid-1980s.

Uranium

The surplus of weapons-grade highly enriched uranium (HEU) led to an agreement between the USA and Russia for some HEU from Russian warheads and military stockpiles to be diluted for delivery to the USA and then used in civil nuclear reactors. Under the 'Megatons to Megawatts' deal signed in 1994, the US Government is purchasing 500 tonnes of weapons-grade HEU over 20 years from Russia for dilution in Russia and sale to US electric utilities, for US$ 12 billion. This acquisition reached its halfway point in 2005, and by early 2012 the total had reached 442 tonnes, with the elimination of 17,700 nuclear warheads.

Weapons-grade HEU is enriched to over 90% U-235 while normal reactor fuel is usually enriched to around 4-5%. To be used in most commercial nuclear reactors, military HEU must therefore be diluted about 25:1 by blending with depleted uranium (mostly U-238, with about 0.25% U-235), natural uranium (0.7% U-235), or re-enriched depleted uranium tails (about 1.5% U-235).

The Russian HEU is being blended down to 4.4% U-235 in Russia, using 1.5% U-235 (enriched tails) for this. The

contracted 500 tonnes of weapons HEU is resulting in just over 15,000 tonnes of low-enriched (4.4%) uranium over the 20 years. This is equivalent to about 137,000 tonnes of natural uranium, about twice annual world demand. The purchase and blending down is being done progressively. Since 2000 the dilution of 30 tonnes per year of military HEU is displacing about 10,000 tonnes of uranium oxide mine production per year, representing about 12% of the world's reactor requirements and about half of the US supply.

Cylinders containing warhead-derived fuel from Russia

The US Government has declared 174 tonnes of highly-enriched uranium (of various enrichments) from its military stockpiles to be surplus to requirements, and most of this has been blended down, eventually to make about 4300 tonnes of reactor fuel. Further releases are likely, and in the short term most of the military uranium is likely to be blended down to 20% U-235, then stored. In this form it is not useable for weapons.

Plutonium

Disarmament will also give rise to some 150-200 tonnes of weapons-grade plutonium. In 2000, the USA and Russia agreed to dispose of 34 tonnes each by 2014. While it was initially proposed to immobilise some of the US portion, the policy now is to fabricate it with uranium oxide as a mixed oxide (MOX) fuel for burning in existing reactors. A plant is under construction in South Carolina for this fuel fabrication, and meanwhile some trial MOX assemblies (made in France from US military plutonium) have been burned in a US reactor. A revised agreement in 2010 allows Russian plutonium to be used in BN-800 fast reactors, and stretches the timeline to 2018.

Europe has a well-developed MOX capacity and Japan started using it in 2009. This suggests that weapons plutonium could be disposed of relatively quickly. Input plutonium would need to be about half reactor-grade and half weapons-grade[3], but using such MOX as 30% of the fuel in one-third of the world's reactor capacity would remove about 15 tonnes of warhead plutonium per year. This would amount to burning around 1500 warheads per year to produce 140 billion kWh of electricity – enough to supply Sweden or nearly half of Italy.

Over 35 reactors in Europe are licensed to use mixed oxide fuel, and 20 French reactors are using it or licensed to use it as 30% of their fuel. Some new reactors will be able to run with full MOX cores.

If all the military plutonium were used in fast neutron reactors in conjunction with the depleted uranium from enrichment plant stockpiles[4], there would be enough to run the world's commercial nuclear electricity programs for several decades without any further uranium mining.

See also: WNA information paper on *Military Warheads as a Source of Nuclear Fuel*.

3.5 THORIUM AS A NUCLEAR FUEL

Most of this book is concerned with nuclear reactors which use uranium as a fuel. However, in future thorium is also likely to be utilised as a fuel for particular reactors. The thorium fuel cycle has some attractive features, and is described further in Section 5.3.

Thorium is about three times as abundant in the Earth's crust as uranium. Australia and India have considerable quantities of thorium, and India has been developing its whole nuclear energy program to make use of it.

Existing neutron efficient reactor designs, such as CANDU, are capable of operating on a thorium fuel cycle, once they are started using a fissile material such as U-235 or Pu-239. Then the thorium (Th-232) nucleus captures a neutron in the reactor to become fissile uranium (U-233), which continues the reaction. However, there are some practical problems with using thorium in this way.

See also: Section 5.3 and WNA information paper on *Thorium*.

3.6 ACCELERATOR-DRIVEN SYSTEMS

The essence of a conventional nuclear reactor is the controlled fission chain reaction of U-235 and Pu-239. This depends on having a surplus of neutrons to keep it going (a U-235 fission requires one neutron input and produces on average

---

3 Reactor-grade plutonium has about one-third non-fissile isotopes; weapons grade is fairly pure Pu-239.
4 When uranium is enriched for a conventional reactor, about seven times more depleted uranium is produced than the enriched product. If uranium is enriched to 93% U-235 for a weapons program about 200 times more depleted uranium than enriched product is produced. All this, some 1.2 million tonnes and comprising a very large proportion of all uranium ever mined, is 'fertile' material and thus potential fast breeder fuel.

Short pulse linear accelerator at Idaho State University, USA. (Photograph supplied by ISU Photographic Services)

2.5 neutrons). However, without such a surplus, a nuclear reaction can be sustained by input of neutrons produced by spallation impact on heavy element targets bombarded by protons in a high-energy accelerator.

If the spallation target is surrounded by a blanket assembly of nuclear fuel, such as fissile isotopes of uranium or plutonium (or thorium which can breed to U-233), there is a possibility of sustaining a fission reaction. This is known as an accelerator-driven system (ADS).

In such a subcritical nuclear reactor, the neutrons produced by spallation would be used to cause fission in the fuel, assisted by further neutrons arising from that fission, though there are insufficient of the latter to sustain a chain reaction. One then has a nuclear reactor which could be turned off simply by stopping the proton beam, rather than needing to insert control rods to absorb neutrons and make the fuel assembly subcritical.

The fuel may be mixed with long-lived fission products or even transuranic nuclides from conventional reactors to incinerate these. Thus the other possible role of a subcritical nuclear reactor or ADS is the destruction of long-lived radioisotopes through transmutation, mainly by fast neutrons. While in principle, the ADS may be able to convert some long-lived transuranic elements into (generally) short-lived fission products and yield some energy in the process, recent research has been focused on fast neutron reactors for this task.

3.7 PHYSICS OF A NUCLEAR REACTOR

Neutrons in motion are the starting point for everything that happens in a nuclear reactor.

When a neutron passes near to a heavy nucleus, for example uranium-235 (U-235), the neutron may be captured by the nucleus and this may or may not be followed by fission. Capture involves the addition of the neutron to the uranium nucleus to form a new compound nucleus. A simple example is U-238 + n → U-239, which represents formation of the nucleus U-239. The new nucleus may decay into a different nuclide. In this example, U-239 becomes Np-239 after emission of a beta particle (electron), and fissile Pu-239 after emission of another one. But in certain cases the initial capture is rapidly followed by the fission of the new nucleus. Whether fission takes place, and indeed whether capture occurs at all, depends on the velocity of the passing neutron and on the particular heavy nucleus involved.

Nuclear fission

Fission may take place in any of the heavy nuclei after capture of a neutron. However, low-energy (slow, or thermal) neutrons are able to cause fission only in those isotopes of uranium and plutonium whose nuclei contain odd numbers of neutrons (e.g. U-233, U-235, and Pu-239). Thermal fission may also occur in some other transuranic elements whose nuclei contain odd numbers of neutrons. For nuclei containing an even number of neutrons, fission can only occur if the incident neutrons have energy above about one million electron volts (MeV). (Newly-created fission neutrons are in this category and move at about 7% of the speed of light, while moderated neutrons move a lot slower, at about eight times the speed of sound.)

A neutron is said to have thermal energy when it has slowed down to be in thermal equilibrium with the surroundings (when the kinetic energy of the neutrons is similar to that possessed by the surrounding atoms due to their random thermal motion). The main

Figure 14. Neutron cross-sections for fission of uranium and plutonium

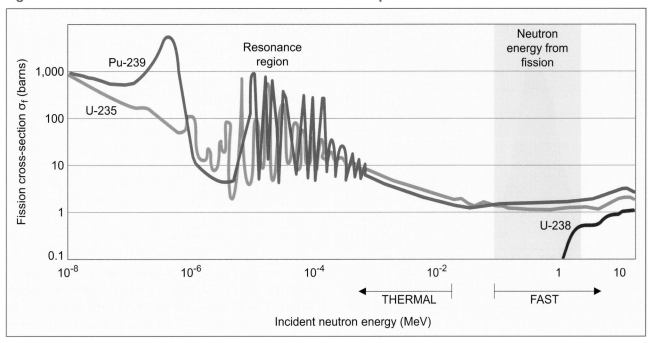

Note that both scales are logarithmic.
Sources: OECD/NEA 1989, Plutonium fuel – an assessment. Taube 1974, Plutonium – a general survey
1 barn = $10^{-28}m^2$, 1 MeV = 1.6 x 10^{-13} J

application of uranium fission is in thermal reactors fuelled by U-235 and incorporating a moderator such as water to slow the neutrons down. The most common examples of this are light water reactors (pressurised water reactors and boiling water reactors).

Other heavy nuclei that are fissile (implying thermal fission) are U-233, Pu-239 and Pu-241. Each of these is produced artificially in a nuclear reactor, from the fertile nuclei Th-232 (in certain reactors), U-238 and Pu-240 respectively. U-235 is the only naturally-occurring isotope which is thermally fissile, and it is present in natural uranium at a concentration of 0.7%. U-238 and Th-232 are the main naturally-occurring fertile isotopes.

The probability that fission or any another neutron-induced reaction will occur is described by the neutron cross-section for that reaction. This may be imagined as an area surrounding the target nucleus and within which the incoming neutron must pass if the reaction is to take place. The fission and other cross-sections increase greatly as the neutron velocity reduces from around 20,000 km/s to 2 km/s, making the likelihood of some interaction greater. In nuclei with an odd number of neutrons, such as U-235, the fission cross-

section becomes very large at the thermal energies of slow neutrons (see Figure 14).

As implied previously, high-energy (> 0.1 MeV) neutrons travel too quickly to have much interaction with the nuclei in the fuel. We therefore say that the fission cross-section of those nuclei is much reduced at high neutron energies relative to its value at thermal energies (for slow neutrons). It is nonetheless possible to use this so-called fast fission in a fast neutron reactor whose design minimises the moderation of the high-energy neutrons produced in the fission process (see next section).

Nuclear fission – the process

Using U-235 in a thermal reactor as an example, when a neutron[5] is captured the total energy is distributed amongst the 236 nucleons (protons & neutrons) now present in the compound nucleus. This nucleus is relatively unstable, and it is likely to break into two fragments of around half the mass each. These fragments are nuclei found around the middle of the Periodic Table and the probabilistic nature of the break-up leads to several hundred possible combinations (see Figure 15). Creation of the fission fragments is followed almost

---

[5] The chain reaction is started by inserting some beryllium mixed with polonium, radium or other alpha-emitter. Alpha particles from the decay cause the release of neutrons from the beryllium as it turns to carbon-12.

Figure 15. Distribution of fission products

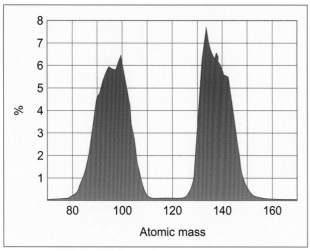

From fuel 65% U, 35% Pu fissions

instantaneously by emission of a number of neutrons (typically 2 or 3, average 2.5), which enable the chain reaction to be sustained.

About 85% of the energy released is initially the kinetic energy of the fission fragments. However, in solid fuel they can only travel a microscopic distance, so their energy becomes converted into heat. The balance of the energy comes from gamma rays emitted during or immediately following the fission process and from the kinetic energy of the neutrons. Some of the latter are immediate (so-called prompt neutrons), but a small proportion (0.7% for U-235, 0.2% for Pu-239) is delayed, as these are associated with the radioactive decay of certain fission products. The longest delayed neutron group has a half-life of about 56 seconds.

The delayed neutron release is the crucial factor enabling a chain reacting system (or reactor) to be controllable and to be able to be held precisely critical. At criticality the chain reacting system is exactly in balance, such that the number of neutrons produced in fissions remains constant. This number of neutrons may be completely accounted for by the sum of those causing further fissions, those otherwise absorbed, and those leaking out of the system. Under these circumstances the power generated by the system remains constant. To raise or lower the power, the balance must be changed (using the control system) so that the number of neutrons present (and hence the rate of power generation) is either reduced or increased. The control system is used to restore the balance when the desired new power level is attained.

The number of neutrons and the specific fission products from any fission event are governed by statistical probability, in that the precise break-up of a single nucleus cannot be predicted. However, conservation laws require the total number of nucleons and the total energy to be conserved. The fission reaction in U-235 produces fission products such as Ba, Kr, Sr, Cs, I and Xe with atomic masses distributed around 95 and 135. Examples may be given of typical reaction products, such as:

U-235 + n → Ba-144 + Kr-90 + 2n + about 200 MeV
U-235 + n → Ba-141 + Kr-92 + 3n + 170 MeV
U-235 + n → Te-139 + Zr-94 + 3n + 197 MeV

In such an equation, the atomic number is conserved, e.g. 235 + 1 = 141 + 92 + 3, but a small loss in atomic mass may be shown to be equivalent to the energy released. Both the barium and krypton isotopes subsequently decay and form more stable isotopes of neodymium and yttrium, with the emission of several electrons from the nucleus (beta decays). It is the beta decays, with some associated gamma rays, which make the fission products highly radioactive. This radioactivity decreases with time.

The total binding energy released in fission varies with the precise break-up, but averages about 200 MeV for U-235 or 3.2×10^{-11} joule[6]. That from Pu-239 is about 210 MeV per fission[6]. (This contrasts with 4 eV or 6.5×10^{-19} J per molecule of carbon dioxide released in the combustion of carbon in fossil fuels.)

About 6% of the heat generated in the reactor core originates from radioactive decay of fission products and transuranic elements formed by neutron capture, mostly the former. This must be allowed for when the reactor is shut down, since heat generation continues after fission stops. It is this decay which makes used fuel initially generate heat and hence need cooling, as very publicly demonstrated in the Fukushima accident when cooling was lost an hour after shutdown and the fuel was still producing about 1.5% of its full-power heat. Even after one year, typical used fuel generates about 10 kW of decay heat per tonne, decreasing to about 1 kW/t after ten years.

Neutron capture: transuranic elements and activation products

Neutrons may be captured by non-fissile nuclei, and some energy is produced by this mechanism in the form of gamma rays as the compound nucleus de-excites. The resultant new nucleus may become more stable by emitting

---

[6] These are total available energy release figures, consisting of kinetic energy values of the fission fragments plus neutron, gamma and delayed energy releases which add about 30 MeV.

alpha or beta particles. Neutron capture by one of the uranium isotopes will form what are called transuranic elements, actinides beyond uranium in the Periodic Table.

Since U-238 is the major proportion of the fuel element material in a thermal reactor, capture of neutrons by U-238 and the creation of U-239 is an important process.

- U-239 quickly emits a beta particle to become neptunium-239.
- Np-239 in turn emits a beta particle to become plutonium-239, which is relatively stable.
- Some Pu-239 nuclei may capture a neutron to become Pu-240, which is less stable.
- By further neutron capture, some Pu-240 nuclei may in turn form Pu-241.
- Pu-241 also undergoes beta decay to americium-241 (the heart of household smoke detectors).

As already noted, Pu-239 is fissile in the same way as U-235, i.e. with thermal neutrons. It is the other main source of energy in any nuclear reactor. If fuel is left in the reactor for a typical three years, about two-thirds of the Pu-239 is fissioned with the U-235, and it contributes about one-third of the energy output. The masses of its fission products are distributed around 100 and 135 atomic mass units.

The main transuranic constituents of used fuel are isotopes of plutonium, curium, neptunium and americium. These are alpha-emitters and have long half-lives, decaying on a similar timescale to the uranium isotopes. They are the reason that used fuel needs secure disposal beyond the few thousand years or so which might be necessary for the decay of fission products alone.

Apart from transuranic elements in the reactor fuel, activation products are formed wherever neutrons impact on any other material surrounding the fuel. Activation products in a reactor (and particularly its steel components exposed to neutrons) range from tritium (H-3) and carbon-14 to cobalt-60, iron-55 and nickel-63. The latter four radioisotopes create difficulties during eventual demolition of the reactor, and affect the extent to which materials can be recycled.

Fast neutron reactors

In a conventional fast neutron reactor the fuel in the core is Pu-239 and the abundant neutrons designed to

The Beloyarsk plant in Russia has the world's largest fast neutron reactor (BN-600) with a larger one (BN-800) under construction

leak from the core breed more Pu-239 in a fertile blanket of U-238 around the core. A minor fraction of U-238 might be subject to fission, but most of the neutrons reaching the U-238 blanket will have lost some of their original energy and are therefore subject only to capture and thus breeding of Pu-239.[7] Cooling of the fast reactor core requires a heat transfer medium which has minimal moderation of the neutrons, and hence liquid metals are used, typically sodium or a mixture of sodium and potassium.

Such reactors are up to 100 times more efficient at converting fertile material than ordinary thermal reactors because of the arrangement of fissile and fertile materials, and there is some advantage from the fact that Pu-239 yields more neutrons per fission than U-235. Although both yield more neutrons per fission when split by fast rather than slow neutrons, this is incidental since the fission cross sections are much smaller at high neutron energies. While the conversion ratio (the ratio of new fissile nuclei to fissioned nuclei) in a normal reactor is around 0.6, that in a fast reactor may exceed 1.0. Fast neutron reactors may be designed as breeders to yield more fissile material than they consume, or to be plutonium burners to dispose of excess plutonium. A plutonium burner would be designed without a breeding blanket, simply with a core optimised for plutonium fuel, and this is the likely shape of future fast neutron reactors, even if they have some breeding function.

Fast reactors have a strong negative temperature coefficient (the reaction slows as the temperature rises unduly), an inherent safety feature, and the basis of automatic load-following in some new designs, by controlling coolant flow.

---

[7] The U-238 is also fissionable when hit by a neutron with over 1 MeV of kinetic energy, but few of the resulting neutrons are energetic enough to produce further U-238 fissions, so no chain reaction is possible due to it alone.

Control of fission

Fission of U-235 nuclei typically releases two or three neutrons, with an average of about 2.5. One of these neutrons is needed to sustain the chain reaction at a steady level of controlled criticality; on average, the others leak from the core region or are absorbed in non-fission reactions. Neutron-absorbing control rods are used to adjust the power output of a reactor. These typically use boron and/or cadmium (both are strong neutron absorbers) and are inserted among the fuel assemblies. When they are slightly withdrawn from their position at criticality, the number of neutrons available for ongoing fission exceeds unity (i.e. criticality is exceeded) and the power level increases. When the power reaches the desired level, the control rods are returned to the critical position and the power stabilises.

The ability to control the chain reaction is entirely due to the presence of the small proportion of delayed neutrons arising from fission. Without these, any change in the critical balance of the chain reaction would lead to a virtually instantaneous and uncontrollable rise or fall in the neutron population. It is also relevant to note that safe design and operation of a reactor sets very strict limits on the extent to which departures from criticality are permitted. These limits are built into the overall design.

While fuel is being burned in the reactor, it is gradually accumulating fission products and transuranic elements which cause additional neutron absorption. The control system has to be adjusted to compensate for the increased absorption. When the fuel has been in the reactor for three years or so, this build-up in absorption, along with the metallurgical changes induced by the constant neutron bombardment of the fuel materials, dictates that the fuel should be replaced. This effectively limits the burn-up to about half of the fissile material, and the fuel assemblies must then be removed and replaced with fresh fuel. Fuel life can be extended by use of burnable poisons such as gadolinium, the effect of which compensates for the build-up of neutron absorbers.

Neutrons released in fission are initially fast (velocity about 10^9 cm/s, or energy above 1 MeV), but fission in U-235 is most readily caused by slow neutrons (velocity about 10^5 cm/s, or energy about 0.02 eV). A moderator material comprising light atoms thus surrounds the fuel

Construction of a new EPR at Flamanville 3 (EDF Photo library/Alexis Morin)

rods in a reactor. Without absorbing too many, it must slow down the neutrons in elastic collisions (compare it with collisions between billiard balls on an atomic scale). In a reactor using natural (unenriched) uranium, the only suitable moderators are graphite and heavy water (these have low levels of unwanted neutron absorption). With enriched uranium (*i.e.* increased concentration of U-235), ordinary (light) water may be used as moderator. (Water is also commonly used as a coolant, to remove the heat and generate steam.)

Other features may be used in different reactor types to control the chain reaction. For instance, a small amount of boron may be added to the cooling water and its concentration reduced progressively as other neutron absorbers build up in the fuel elements. (For emergency situations, provision may be made for rapidly adding an excessive quantity of boron to the water.)

Commercial power reactors are usually designed to have negative temperature and void coefficients. The significance of this is that if the temperature should rise beyond its normal operating level, or if boiling should occur beyond an acceptable level, the balance of the chain reaction is affected so as to reduce the rate of fission and hence reduce the temperature. One mechanism involved is the Doppler effect[8], whereby U-238 absorbs more neutrons as the temperature rises, thereby pushing the neutron balance towards subcritical. Another important mechanism, in light water reactors, is that the formation of steam within the water moderator will reduce its density and hence its moderating effect, and this again will tilt the neutron balance towards subcritical.

Extending fission

In naval reactors used for propulsion, where fuel changes are inconvenient, the fuel is enriched to higher levels initially and burnable poisons – neutron absorbers – are incorporated. Hence as the fission products and transuranic elements accumulate, the 'poison' is depleted and the two effects tend to cancel one another out. To maximise the burn-up of commercial reactor fuel, burnable poisons such as gadolinium are increasingly used, along with increasing enrichment towards 5% U-235.

Another major development gathering impetus is towards burning transuranic elements (plutonium and minor actinides) which are extracted from used fuel. If these remain in discarded used fuel or in wastes from reprocessing used fuel, they will increase the heat load in repositories and complicate the task of designing them. Once the best way of reprocessing used fuel to recover them is resolved, the issue is: how best to fission them? The answer appears to be: in fast neutron reactors, since these maximise fission, even yielding some energy. But various thermal reactor and accelerator system methods are also being investigated. (A related question is how to transmute the longer-lived fission products such as technetium-99, iodine-129 and caesium-135 into shorter-lived ones. Here, neutron capture is the objective, and a liberal supply of slow neutrons is required.)

---

[8] As the temperature rises, the movement of atoms (here: mostly U-238) increases, which in turn increases the likelihood that they will be moving away from fast neutrons, thus diminishing relative velocity and greatly increasing the chance of capture.

4. Types of nuclear power reactor

4.1 TODAY'S POWER REACTORS

Several generations of reactors are commonly distinguished. The first generation of reactors was developed in 1950-60s and, outside the UK, none are still running. The next generation of power reactors is typified by the present US fleet and throughout Europe, as well as most of those in operation elsewhere. Nearly all of the 436 nuclear power reactors currently operable around the world are second-generation designs, which have proved to be safe and reliable, but they are being superseded by more advanced designs.

Loviisa nuclear power plant, Finland

Table 8. Nuclear power plants commercially operable

Reactor type	Main countries	Number	GWe	Fuel	Coolant	Moderator
Pressurised water reactor (PWR)	US, France, Japan, Russia, China	271	270.4	Enriched UO_2	Water	Water
Boiling water reactor (BWR)	US, Japan, Sweden	84	81.2	Enriched UO_2	Water	Water
Pressurised heavy water reactor, *e.g.* 'CANDU' (PHWR)	Canada, India	48	27.1	Natural UO_2	Heavy water	Heavy water
Gas-cooled reactor (AGR & Magnox)	UK	17	9.6	Natural U (metal); enriched UO_2	CO_2	Graphite
Light water graphite reactor (RBMK & EGP)	Russia	11 + 4	10.4	Enriched UO_2	Water	Graphite
Fast neutron reactor (FNR)	Russia	1	0.6	PuO_2 and UO_2	Liquid sodium	None
Total		**436**	**399.3**			

Descriptions of these different kinds of reactors may be found in the WNA information paper on Nuclear Power Reactors.
GWe = capacity in gigawatts (thousands of megawatts).

4.2 ADVANCED POWER REACTORS

The first few advanced reactors are operating in Japan and others are under construction or ready to be ordered. These are sometimes called Generation III types, and are improved versions of those now operating, but there is no clear consensus on the transition point. Generation IV designs are still on the drawing board and will not be commercialised before the 2030s.

Qinshan Phase III nuclear power plant (CANDU), China (CNNC)

Reactor suppliers in North America, Japan, Europe and Russia and elsewhere have several new nuclear reactor designs either approved or at advanced stages of planning, and others at a research and development stage. These incorporate safety improvements and will also be simpler to build, operate, inspect and maintain, thus increasing their overall reliability and economy.

In general, the new generation reactors have the following characteristics:

- A standardised design for each type to expedite licensing, reduce capital cost and reduce construction time.
- A simpler and more rugged design, making them easier to operate and less vulnerable to operational upsets.
- Higher availability and longer operating life – typically 60 years, as against 40 or so for earlier designs (extendable with capital investment).
- Reduced possibility of accidents in which the reactor's core melts, particularly through coping with decay heat following shut-down (the problem at Fukushima).
- Resistance of the structure to the serious damage that would allow radiological release from an aircraft impact.
- Higher burn-up of fuel, to use it more fully and efficiently and to reduce the amount of radioactive waste created.

The greatest change from most designs now operating is that many new nuclear plants will have more 'passive' safety features, which rely on gravity, natural convection, etc., requiring no active controls or operational intervention to avoid accidents in the event of malfunction. They will allow operators more time to remedy problems, and provide greater assurance regarding containment of radioactivity in all circumstances.

A separate line of development epitomising this is of small high-temperature reactors with refractory fuel capable of withstanding very high temperatures, and cooled by helium. These are put forward as intrinsically safe, in that no emergency cooling system is needed, and in the event of a problem the units can be left to themselves. Being relatively small, the high surface to volume ratio of the reactor enables dissipation of heat naturally (see also Section 4.5).

Certification of designs is on a national basis, and is safety-based. In Europe, and to some extent more widely, there are moves towards harmonised requirements for reactor design licensing, and the Fukushima accident has sharpened the focus on this need. In the USA there is a design certification process which has approved four advanced reactor designs and is processing three more.

Control room at Balakovo nuclear power plant, Russia

Unit 1 of the Novovoronezh II nuclear power plant, Russia (photo: Hullernuc)

These US Nuclear Regulatory Commission (NRC) approvals mean that as a result of an exhaustive public process, safety issues within the scope of the certified designs have been fully resolved and hence will not be open to legal challenge during licensing for particular plants. Furthermore, utilities will be able to obtain a single NRC licence to both construct and operate a reactor in the USA before construction begins.

Table 9. Main new-generation reactors

Areva **EPR**	1700 MWe, being built in Finland, France, China and soon in UK
Westinghouse **AP1000**	1200 MWe, being built in China, adopted as main design there
GE-Hitachi/ Toshiba **ABWR**	1350 MWe, operating in Japan, being built there and in Taiwan, soon in Lithuania
Gidropress **VVER-1200**	1200 MWe, being built in Russia, soon in Turkey and Belarus
Korea Hydro & Nuclear Power **APR-1400**	1450 MWe, being built in South Korea, soon in the UAE
Mitsubishi **APWR**	1500 MWe, planned for Japan and USA
GE-Hitachi **ESBWR**	1600 MWe, planned for USA

ABWR

The first new-generation design, approved in 1997, was the 1350 MWe advanced boiling water reactor (ABWR), examples of which are in commercial operation in Japan, with more under construction there and in Taiwan. One is planned for Lithuania, and two were proposed for South Texas, USA. In awarding final design certification the US Nuclear Regulatory Commission (NRC) said that it exceeded NRC "safety goals by several orders of magnitude". It is marketed by both Toshiba and GE-Hitachi.

AP1000

In 2005, the NRC gave design certification to the 1200 MWe Westinghouse AP1000, which was the first late third-generation design, representing a further step forward in safety. It is capable of running on a full mixed oxide fuel (MOX) core if required, and its modular design potentially reduces construction time to 36 months. The first ones are under construction at Sanmen and Haiyang in China, where many more are planned. Four are being built in the USA to come on line from 2016,

and six more are planned there. China is developing a CAP-1400 based on it.

EPR/US-EPR

Areva has developed a large (1600 and up to 1750 MWe) European Pressurized Water Reactor (EPR), which is the new standard design for France and received French design approval in 2004. It will operate flexibly to follow loads, and is capable of using a full core load of MOX. The first EPR unit is being built at Olkiluoto in Finland, the second at Flamanville in France, the third European one will be at Penly in France, and a further two are under construction at Taishan in China. Several are planned for the UK. A US version, the US-EPR, is undergoing US design certification and the first unit was proposed at Calvert Cliffs.

APWR/US-APWR

Mitsubishi's large 1600 MWe APWR – advanced PWR – was developed in collaboration with four utilities and the first two are planned for Tsuruga in Japan. It will be the basis for the next generation of Japanese PWRs, but in the meantime a US version of 1700 MWe is proposed for construction in two states.

APR-1400

In South Korea, the APR-1400 advanced PWR design has evolved from US origins. Korean design certification was awarded in 2003, and the first of these 1450 MWe reactors are under construction at Shin-Kori, with a total of eight planned at two sites. Four have been sold to the United Arab Emirates, with construction planned to start in 2012.

VVER-1200

From Russia, Gidropress late models of the well-proven VVER-1000 (in AES-91 and AES-92 power plants) units with enhanced safety have been built in China and India. Another is planned for Kozloduy in Bulgaria. Developed from these is a new-generation VVER-1200 reactor (in AES-2006 plants) with longer life, more power and greater efficiency. The lead units are being built at Novovoronezh II and Leningrad II, and others are planned for Akkuyu in Turkey and in Belarus. An evolutionary version of the VVER-1200, known as the VVER-TOI (AES-2010), is due to be submitted for European Utilities Requirements (EUR) accreditation by the end of 2012.

ESBWR

GE-Hitachi has developed the ESBWR of 1550 MWe with passive safety systems, from its ABWR design. It is expected to complete US design certification in 2012 and is proposed for construction in the USA.

Atmea1

This has been developed by the Atmea joint venture established in 2007 by Areva and Mitsubishi Heavy Industries to produce an evolutionary 1150 MWe reactor using the same steam generators as the EPR. This has 60-year life, and the capacity to use mixed-oxide fuel only. It has load-following and frequency control capability. The French regulator approved the general design in 2012. It will be marketed primarily to countries embarking upon nuclear power programs. It has three active and passive redundant safety systems and an additional backup cooling chain, similar to EPR.

Kerena

Together with German utilities and safety authorities, Areva has developed an evolutionary BWR design, the Kerena, of 1290 MWe capacity and 60-year design life. It has two redundant active safety systems and two passive safety systems, with a core catcher, similar to EPR. The reactor is simpler overall and uses high-burnup fuels, giving it refuelling intervals of up to 24 months.

EC6 and ACR

The Canadian Enhanced CANDU-6 (EC6) is a development of earlier CANDU reactors with power increased to 750 MWe gross and flexible fuel options, plus 60-year plant life (with mid-life pressure tube replacement). This is under consideration for new build in Ontario and Romania. A further development, the Advanced Candu Reactor (ACR) design, retains the low-pressure heavy water moderator of its predecessors, but adopts light water cooling and a more compact core. The ACR-1000 of 1200 MWe will run on very low-enriched uranium.

AHWR

India is developing the Advanced Heavy Water Reactor (AHWR) as the third stage in its plan to utilise thorium to fuel its overall nuclear power program. The AHWR is a 300 MWe reactor design moderated by heavy water at low pressure. It is designed to be self-sustaining in relation to U-233 bred from Th-232 (see Section 5.3).

Several other significant advanced reactor designs are proceeding.

See also WNA information papers on *Advanced Nuclear Power Reactors* and *Small Nuclear Power Reactors*. Beyond these, see WNA information paper on *Generation IV Nuclear Reactors*.

4.3 FLOATING NUCLEAR POWER PLANTS

Russia's floating nuclear power plant concept employs a pair of pressurised water reactors similar to those used in icebreakers.

The first of these floating nuclear power plants is under construction in the Baltic shipyard at St Petersburg. The *Akademik Lomonosov* will be located near Vilyuchinsk, in Russia's far east, from about 2014. The two KLT-40S reactors well proven in icebreakers and now using low-enriched fuel (less than 20% U-235) are from OKBM Afrikantov and are mounted on a 21,500 tonne barge. Each reactor has a capacity of 150 MW thermal, or about 35 MWe as well as up to 35 MW of heat for desalination or district heating. Refuelling interval is 3-4 years on site, and at the end of a 12-year operating cycle the whole plant is returned to a shipyard for a two-year overhaul and storage of used fuel, before being returned to service.

4.4 MODULAR LIGHT WATER REACTORS

Modular reactors are small units, usually designed to be built in factories, and intended to be set up as components of a power plant which has several of them operating together. (Modular construction refers to large elements of a power plant, *e.g.* AP1000, being prefabricated and then assembled on site.) Most of these have integral steam generators, *i.e.* inside the pressure vessel. None is yet built.

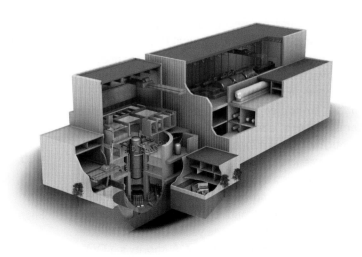

Representation of the ESBWR (GE-Hitachi)

Westinghouse SMR

This is a 225 MWe class integral PWR with passive safety systems and reactor internals including fuel assemblies based closely on those in the AP1000. The steam generator is above the core. The reactor vessel module is 25 metres high and 3.5 metres diameter and will be factory-made and shipped to site by rail, then installed below ground level in a containment vessel. Its passive safety features mean no operator intervention is required for seven days in the event of an accident. The US Department of Energy (DOE) sees this as a "near-term LWR design." Westinghouse has teamed up with utility Ameren Missouri to develop the design and proceed towards building a five-unit plant at Callaway in Missouri.

The first floating nuclear power plant, Akademik Lomonosov, is to be located near Vilyuchinsk, in Russia's far east

mPower

Babcock & Wilcox (B&W) has set up a subsidiary with engineering input from Bechtel to develop the mPower modular reactor, a 125 MWe integral PWR designed to be factory-made and railed to site. The reactor pressure vessel containing core and steam generator is only 3.6 metres diameter. It will be installed below ground level, have an air-cooled condenser, and passive safety systems. It has a "conventional core and standard fuel" with burnable poisons, to give a five-year operating cycle between refuelling, which will involve replacing the entire core as a single cartridge. Several units would be combined into a power station of any size, but most likely 500-750 MWe and using 250 MWe turbine generators, constructed in three years. Up to six modules may be built at Tennessee Valley Authority's (TVA's) Clinch River site in Tennessee, and the DOE sees it as a "near-term LWR design".

NuScale

A smaller unit is the NuScale 45 MWe integral PWR, now involving engineering support from Fluor. It will be factory-built with a three-metre diameter pressure vessel and uses convection cooling. It uses standard design but half-length PWR fuel assemblies. The containment vessel module contains the reactor and steam generator and is installed in a water-filled pool below ground level. A standard power plant would have 12 modules together, giving 540 MWe of capacity. The US DOE sees this as a "near-term LWR design" and in March 2012 it signed an agreement with NuScale regarding constructing a demonstration unit at its Savannah River site in South Carolina.

Holtec SMR

Holtec International has set up a subsidiary to commercialize a 140 MWe reactor concept called Holtec Inherently Safe Modular Underground Reactor (HI-SMUR 140). This is a PWR with external steam generator, using fuel similar to that in larger PWRs, though contained in a single cartridge. It has full passive cooling in operation and after shutdown. The whole reactor system will be installed below ground level. The Shaw Group is providing engineering support for the design. In March 2012 the US DOE signed an agreement with Holtec regarding constructing a demonstration 160 MWe unit at Savannah River in South Carolina. The unit was then designated SMR-160, "based on HI-SMUR".

Several other small reactors are being developed, though not specifically as modular types. They include CAREM (27 MWe) in Argentina, SMART (100 MWe) in South Korea, and ACP100 (100-150 MWe) in China.

4.5 HIGH TEMPERATURE REACTORS

High-temperature gas-cooled reactors (HTR) are modular reactors, limited in size by their design. They use graphite as moderator (unless fast neutron type) and either helium, carbon dioxide or possibly nitrogen as primary coolant.

Building on the experience of several innovative reactors built and operated in the 1960s to 1980s, especially in Germany, new HTRs are being developed which are relatively small and hence designed to be built as multiple modules. They will be capable of delivering high-temperature helium (eventually up to 950°C) either for industrial heat application or directly to drive gas turbines for electricity (the Brayton cycle) with about 48% thermal efficiency possible.[1] Technology developed in the last decade makes HTRs more practical than in the past, though the direct cycle will be a further technological step which means that there must be high integrity of fuel and reactor components.

Fuel for these reactors is in the form of particles less than a millimetre in diameter. Each has a kernel of uranium oxycarbide, with the uranium enriched up to 17% U-235. This is surrounded by layers of carbon and silicon carbide, giving a containment for fission products which is stable to 1600°C or more. There are two ways in which these particles are arranged: in blocks – hexagonal prisms of graphite; or in 'pebbles' of graphite encased in silicon carbide, each the size of a billiard ball and containing about 15,000 fuel particles and 9 grams of uranium enriched to 9-10% U-235.

HTR performance includes great flexibility in loads (ranging from 40-100% of capacity) without loss of thermal efficiency, and with rapid change in power settings. Power density in the core is about one-tenth of that in light water reactors, and if coolant circulation ceases the fuel will survive initial high temperatures while the reactor shuts itself down – giving a high level of inherent safety. Power control is by varying the coolant pressure and hence flow, relying on a strong negative temperature coefficient (whereby fission slows as temperature rises).

HTRs can potentially use thorium-based fuels, such as high-enriched uranium with thorium, U-233 with thorium, and plutonium with thorium. Much of the experience with thorium fuels has been in HTRs.

HTR-PM

The HTR-10, a small pebble bed reactor at China's Tsinghua University's Institute of Nuclear Energy Technology, has been running successfully since 2000. A larger version, with twin reactors driving a single steam cycle turbine is under construction at Shidaowan, in Shandong Province. This demonstration HTR-PM plant, with a total capacity of 210 MWe from twin reactors, is to pave the way for an 18-module full-scale power plant on the same site, also using the steam cycle. Plant life is envisaged as 60 years with 85% load factor. Huaneng, one of China's major generators, is the lead organisation involved. The HTR-PM rationale is both eventually to replace conventional reactor technology for power, and also to provide for future hydrogen production.

EM[2]

General Atomics is developing its Energy Multiplier Module (EM[2]) design, based on its earlier Gas Turbine Modular Helium Reactor (GT-MHR) design. EM[2] is a 240 MWe helium-cooled fast-neutron HTR operating at 850°C and fuelled with used PWR fuel or depleted uranium, plus some low-enriched uranium as starter. Used fuel from this will be processed to remove fission products and the balance recycled as fuel for subsequent rounds, each time topped up with a few tonnes of further used PWR fuel. Each refuelling cycle may be as long as

Table 10. High temperature reactors

Country & developer	Reactor	Module size	Temperature °C	Design progress
China (INET)	HTR-PM	105 MWe	750	Demo plant under construction
France (Areva)	SC-HTGR	250 MWe	750	Under development for USA
USA (General Atomics)	EM[2]	240 MWe	850	Under development

---

[1] In the short term early demonstration plants will use the steam cycle for electricity generation.

30 years. With repeated recycling the amount of original natural uranium (before use by PWR) used goes up from 0.5% to 50% at about cycle 12. A 48% thermal efficiency is claimed, using the Brayton cycle. EM² would also be suitable for process heat applications. The main pressure vessel can be trucked or railed to site, and installed below ground level.

Antares/Areva SC-HTGR

Another full-size HTR design is being put forward by Areva. It is based on the GT-MHR and has also involved Fuji. Reference design is 625 MW thermal with prismatic block fuel like the GT-MHR. Core outlet temperature is 750°C for the steam-cycle HTR version (SC-HTGR), though an eventual very high temperature reactor (VHTR) version is envisaged with 1000°C and direct cycle. The present concept uses an indirect cycle, with steam in the secondary system, or possibly a helium-nitrogen mix for VHTR. It was selected in 2012 for the US Next Generation Nuclear Plant, with two-loop secondary steam cycle, the 625 MWt probably giving 250 MWe per unit, but the primary focus being the 750°C helium outlet temperature for industrial application.

4.6 FAST NEUTRON REACTORS

Fast neutron reactors (FNR) are a different technology from those considered so far. They generate power from plutonium by much more fully utilising the uranium-238 in the reactor fuel assembly, instead of being fuelled with just the fissile U-235 isotope used in most reactors (see also Section 3.7). If they are designed to produce more plutonium than they consume, they are called fast breeder reactors (FBR). But many designs are net consumers of fissile material including plutonium.[2] Fast neutron reactors also can burn long-lived actinides which are separated from used fuel out of ordinary reactors.

For many years the focus was on the potential of this kind of reactor to produce more fuel than it consumed, but with low uranium prices (mid-1980s to about 2003), the interest waned. However, several countries maintain research and development programs for FBRs, which are, generically, fast neutron reactors, and over 300 reactor-years of operating experience has been gained on this type of plant. Some earlier programs faltered,

and significant technical and materials problems were encountered. The French program was derailed by political decision, the UK operated two units over more than 30 years, the Japanese program was suspended due to a coolant leak, and only the Russian program continues with any vigour (see Table 11), though China is catching up.

About 20 FNRs have already been operating, some since the 1950s, mostly as FBRs.

> **The fast neutron reactor has the potential for utilising virtually all of the uranium produced from mining operations.**

As described in Chapter 3, overall about 60 times more energy can be extracted from the original uranium by the fast breeder cycle than can be produced by the current light water reactors operating in 'open cycle'. This extremely high energy efficiency makes the breeder an attractive energy conversion system. However, high capital costs and an abundance of relatively low-cost uranium means that they are generally not competitive at present.

For this reason there was little interest in fast reactors until recently. The 1250 MWe French Superphenix FBR operated 1985-98 before being closed by political edict. The first of four 500 MWe Indian FBRs is under

Monju fast neutron reactor, Japan (JNC)

---

[2] If the ratio of final to initial fissile content is less than 1 they are burners, consuming more fissile material (U-235, Pu and minor actinides) than they produce (fissile Pu); if more than 1 they are breeders. This is the burn ratio or breeding ratio.

construction, to pave the way to greater use of thorium as a fuel, at Kalpakkam. Japan's Monju prototype commercial FBR was connected to the grid in August 1995 but has been plagued by incidents. China's Experimental Fast Reactor started up in mid-2010 and a 1000 MWe version is planned based on this.

The fast neutron reactor has no moderator and uses plutonium as its basic fuel since that fissions sufficiently with fast neutrons to keep going. At the same time the number of neutrons produced per fission is more than from uranium, and this means that there are enough (after losses) not only to maintain the chain reaction but also continually to convert some of the abundant U-238 in depleted or recycled uranium into fissile plutonium. Also, the fast neutrons are more efficient than slow ones in doing this breeding. These are the main reasons for avoiding the use of a moderator.

Conventional fast reactor designs have depleted uranium – basically U-238 – comprising a 'fertile blanket' around the core, and it is this which is reprocessed to recover plutonium. However, fast reactor concepts being developed for the Generation IV program will simply have a core so that the plutonium production and consumption both occur there. Either way, the fast reactor 'burns' and can 'breed' fissile plutonium.[3] Depending on the design, it is possible to recover from reprocessing the used fuel enough fissile plutonium for the reactor's own needs, with some left over for future breeder reactors or for use in conventional reactors as MOX fuel. Conceptually, refuelling means simply adding a little natural or depleted uranium – about one or two percent of the total required for a comparable light water reactor. The fast reactor fuel cycle is illustrated in Figure 18 on page 59.

Table 11. Some fast neutron reactors – past & present

Country	Name	Output	Full operation
USA	EBR-II	20 MWe	1963-94
	Fermi 1	66 MWe	1963-72
	Fast Flux Test Facility	400 MWt	1980-93
UK	Dounreay FR	15 MWe	1959-77
	Prototype FR	270 MWe	1974-94
France	Rapsodie	40 MWt	1966-82
	Phenix	250 MWe	1973-2009
	Superphenix	1240 MWe	1985-98
Germany	KNK 2	21 MWe	1977-91
India	FBTR	40 MWt	1985-
Japan	Joyo	140 MWt	1978-
	Monju	280 MWe	1994-96, 2010
Kazakhstan	BN-350	135 MWe	1972-99
Russia	BR 5/10	5, then 8 MWt	1959-71, 1973-
	BOR-60	10 MWe	1969-
	BN-600	600 MWe	1980-

Those which produce or produced electricity are shown with MWe capacity, the others MW thermal.
Only the BN-600 is in commercial operation. Monju is expected to restart in 2012.

---

[3] Both U-238 and Pu-240 are 'fertile' (materials), *i.e.* by capturing a neutron they become (directly or indirectly) fissile Pu-239 and Pu-241 respectively.

Fast neutron reactors have a high thermal efficiency due to their high-temperature operation. They also have an even higher power density than conventional ones and are normally cooled by liquid metal such as sodium, lead, or lead-bismuth, with high conductivity and boiling point and no moderating effect. They operate at around 500-550°C (instead of around 325°C), at or near atmospheric pressure. Although in many ways liquid metal coolant is difficult to handle chemically, in some respects it is more benign overall than very high pressure water, which requires robust engineering on account of the pressure. Experiments on a 19-year old UK breeder reactor before it was decommissioned in 1977 showed that the liquid sodium cooling system made it less sensitive to coolant failures than the more conventional very high pressure water and steam systems in light water reactors. More recent operating experience with large French and UK prototypes has confirmed this.

There is renewed interest in fast reactors due to their ability to fission actinides recovered from ordinary reactor used fuel. The fast neutron environment minimises neutron capture reactions and maximises fissions in actinides. This means fewer long-lived nuclides in high-level wastes (the fission products being preferable due to shorter half-lives). But as well as this incinerator role, there is the potential for using as fuel the 1.5 million tonnes of depleted uranium now stockpiled.

Longer-term, FNRs are expected to be the main technology in use, and national policies in several countries affirm this. Five of the seven reactor designs being developed by the Generation IV International Forum[4] are FNRs.

BN series

Russia's BN-350 FBR operated in Kazakhstan for 27 years and about half of its output was used for water desalination. The BN-600 FBR at Beloyarsk has been supplying electricity to the Russian grid since 1981 and is said to have the best operating and production record of all Russia's nuclear power units. The BN-800 fast reactor nearing completion there is designed to supersede the BN-600 unit and utilise MOX fuel with both reactor-grade and weapons plutonium. The success of BN-600 represents a technological advantage for Russia and has significant export or collaborative potential with Japan and China. Further BN-800 units are planned in Russia, and in 2009 the design was sold to China, with the first two to be built from 2013.

SVBR

A smaller and newer Russian design is the Lead-Bismuth Fast Reactor (SVBR) of 100 MWe, from Gidropress. This is an integral design, with the steam generators sitting in the same Pb-Bi pool at about 490°C as the reactor core. It is designed to use a wide variety of fuels, though the demonstration unit will initially use uranium enriched to 16.3%. With U-Pu MOX fuel it would operate in closed cycle. Refuelling interval is 7-8 years. The SVBR-100 unit would be factory-made and shipped as a 4.5m diameter, 7.5m high module, then installed in a tank of water to give passive heat removal and shielding. A power station with 16 such modules is expected to supply electricity at lower cost than any other new Russian technology as well as achieving inherent safety and high proliferation resistance. AKME Engineering plans to complete the design development and put on line a 100 MWe pilot facility by 2017 at Dimitrovgrad. The SVBR-100 could be the first reactor cooled by heavy metal to generate electricity. It is a multi-function reactor, for power, heat or desalination.

PRISM

PRISM (Power Reactor Innovative Small Module) is a GE-Hitachi (GEH) compact modular pool-type reactor with passive cooling for decay heat removal. After 30 years of development it represents GEH's advanced solution to closing the fuel cycle in the USA. A PRISM power block consists of two modules of 311 MWe each, operating at over 500°C. The modules below ground level contain the complete primary system with sodium coolant. The metal plutonium and depleted uranium fuel is obtained

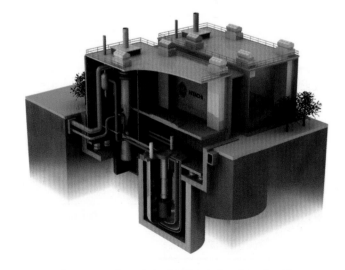

Representation of the PRISM design (GE-Hitachi)

4 The Generation IV International Forum (GIF) is an international collective representing governments of 13 countries collaboratively developing seven (originally six) reactor designs for deployment in the 2030s.

from used light water reactor fuel and incorporates minor actinides with the plutonium. Fuel stays in the reactor about six years, with one-third removed every two years. Used PRISM fuel is recycled after removal of fission products, though not necessarily into PRISM units. The UK government is considering a proposal to use PRISM technology to dispose of the UK's plutonium stockpile.

See also WNA information paper *Fast Neutron Reactors*, and relevant sections of *Advanced Nuclear Power Reactors* and *Small Nuclear Power Reactors*.

4.7 VERY SMALL NUCLEAR POWER REACTORS

Among several very small nuclear power reactor projects with prospects of deployment in the next 10-15 years, three are noteworthy. All are to be factory-built and may comprise modules in a larger plant or be stand-alone units, for either power or heat. Initially all three are fast neutron reactors. All have a sealed core and will be shipped to site and installed below ground level. They operate at atmospheric pressure, and their passive safety features and automatic load following is achieved due to the reactivity feedback – constrained coolant flow leads to higher core temperature, which slows the reaction. All are designed for use in developing countries and are intrinsically resistant to misuse.

STAR

The largest of the three is the Secure Transportable Autonomous Reactor (STAR), a modular fast neutron reactor with passive safety features and cooled by lead-bismuth eutectic.[5] It is a 300-400 MW thermal (about 150 MWe) fast reactor being developed by Argonne National Laboratory in the USA. The 2.5 m diameter reactor module and the steam generators sit inside the primary coolant which circulates by natural convection. After a 15-year life without refuelling, the whole reactor module cartridge is removed from the coolant and sent to a regional centre for recycling the nitride fuel.

A smaller STAR variant is the 20 MWe SSTAR being developed by Toshiba and others in Japan as part of the Generation IV program. It will be a sealed unit with the core one metre high and 1.2 m diameter, and a 20-year life. The whole reactor would then be removed for recycling the fuel.

4S 'nuclear battery'

The Super-Safe, Small and Simple (4S) system is being developed by Toshiba in Japan. It uses well-proven sodium as coolant (with electromagnetic pumps) and has passive safety features. The 10 MWe fast reactor unit would drive a steam cycle and be capable of three decades of continuous operation without refuelling. Metallic alloy fuel is enriched to less than 20% U-235. Steady power output over the core lifetime is achieved by progressively moving upwards an annular reflector around the slender core (0.7m diameter, 2m high). After 14 years a neutron absorber at the centre of the core is removed and the reflector repeats its slow movement up the core for 16 more years. In the event of power loss the reflector falls to the bottom of the reactor vessel, slowing the reaction, and external air circulation gives decay heat removal.

The design has gained considerable support in Alaska due to likely cost being very competitive with diesel in many locations. The town of Galena invited Toshiba to build a 4S reactor in that remote location. A pre-application licensing review is being sought with a view to a demonstration unit being built. Its design is sufficiently similar to an earlier liquid metal-cooled inherently-safe reactor design which went part-way through the US approval process for it to have good prospects of licensing.

Gen4 Module (formerly Hyperion Power Module)

This is a small self-regulating hydrogen-moderated and lead-bismuth-cooled fast reactor of 25 MWe capacity fuelled by 20% enriched uranium nitride. It operates at about 550°C and is designed to operate for seven to ten years before being returned to the factory for refuelling. It is about 1.5 metres wide and 2.5 metres high, so easily portable, and installed below ground level. It is sealed and has no moving parts. A US design certification application is planned for 2012, when the company plans to begin manufacturing the plants in New Mexico. In March 2012, the US DOE signed an agreement with Gen4 Energy for construction of a demonstration unit at its Savannah River site in South Carolina.

See also WNA information paper *Small Nuclear Power Reactors*.

---

[5] *i.e.* with low melting point (125°C) instead of 327°C for lead.

5. The 'front end' of the nuclear fuel cycle

5.1 MINING AND MILLING OF URANIUM ORE

Mining and milling of uranium ore is in most respects the same as that for other metallic minerals. The main difference, which may not be great, is in respect to radioactivity.

Because uranium decays radiologically over geological time, uranium minerals are always associated with other elements such as radium and radon in the radioactive decay series of uranium (see Appendix 2). Therefore, although uranium itself is barely radioactive, the ore which is mined must be regarded as potentially hazardous, especially if it is high-grade ore. The radiation hazards involved are similar to those in many mineral sands mining and treatment operations.

Many of the world's uranium mines have been open-cut and therefore naturally well ventilated. Any underground mine must have an effective ventilation system, and for a uranium mine this is even more important.

Ore grades at most mines worldwide are less than 0.5% uranium oxide (U_3O_8). The Olympic Dam underground mine in Australia, located in the largest known uranium orebody in the world, has ore grade less than 0.1%

U_3O_8. By contrast, Canada's McArthur River and Cigar Lake mines have very high-grade ore and therefore require special remote-control techniques for mining.

The mined ore (*i.e.* rock containing economically recoverable concentrations of uranium) is crushed and ground. The resulting slurry is then leached, usually with sulfuric acid, to dissolve the uranium. The solids remaining after the uranium is extracted are known as tailings. They are pumped as a slurry to the tailings dam, which is engineered to retain them securely. The tailings contain most of the radioactive material in the ore, such as radium.

Some newer mines are in situ leaching (ISL) operations, with recovery of the uranium from the sandy ore taking place underground in an aquifer. Water from the deposit is made slightly acidic and heavily oxygenated, then circulated through the ore via boreholes. The dissolved uranium is extracted in plant at the surface, with the liquor being recirculated.

In each case, conventional mining or ISL, the leach liquor carrying the dissolved uranium goes through a solvent extraction or ion exchange process, followed by precipitation to remove the uranium as a bright yellow precipitate ('yellowcake'). After high-temperature drying, the uranium oxide (U_3O_8), now khaki in colour, is packed into 200-litre drums for shipment. The radiation level one metre from such a drum of freshly processed U_3O_8 is about half that (from cosmic rays) received by a person on a commercial jet flight.

Radiation protection

In Australia all these operations are undertaken under the Australian *Code of Practice & Safety Guide: Radiation Protection and Radioactive Waste Management in Mining and Mineral Processing* (2005), administered by state governments. In Canada, the Canadian Nuclear Safety Commission regulations apply. In both countries, there are strict health standards for gamma radiation and radon gas exposure[1] as well as for ingestion and inhalation of radioactive materials. In other countries, similar regulations

Crushing and grinding section of Ranger mill, northern Australia (Energy Resources of Australia)

Ranger uranium open-cut mine and treatment plant in northern Australia (Energy Resources of Australia)

are in place. Standards apply to both workers and members of the public.

The gamma radiation from uranium ore comes principally from isotopes of bismuth and lead in the uranium decay series. The radon gas emanates from the rock (or tailings) as radium decays[2]. It then decays itself to (solid) radon daughters, which are energetic alpha-emitters. Radon occurs in most rocks and traces

of it are in the air we all breathe. However, at high concentrations it is a health hazard since its short half-life means that disintegrations giving off alpha particles are occurring relatively frequently. Alpha particles discharged in the lung can later give rise to lung cancer.

A number of precautions are taken at any mine to protect the health of workers, and those at a uranium mine are slightly greater:

- Dust is controlled, so as to minimise inhalation of silica, and also gamma- or alpha-emitting minerals. In practice, dust is the main source of radiation exposure in a uranium mine. At the Ranger mine in Australia, it typically contributes up to 2 mSv/yr to a worker's annual dose[3] (see also Table 16 on page 96).
- Radiation exposure of workers in the mine, plant and tailings areas is controlled. In practice, direct radiation levels from the ore and tailings at most mines are usually so low that it would be difficult for a worker to come anywhere near the allowable annual dose. However, in some Canadian mines the dose would be so high that people are excluded from the workings.
- Radon daughter exposure is low in an open cut mine because there is sufficient natural ventilation, and the radon level seldom exceeds 1% of the levels allowable for continuous occupational exposure. In an

McArthur River underground mine – the world's largest high-grade uranium deposit (Cameco)

---

[1] 20 mSv/yr averaged over five years is the maximum allowable radiation dose rate for workers, including radon (and radon daughters) dose. This is in addition to natural background, and excludes medical exposure. See also Appendix 1 and Glossary for definitions.

[2] Radon here normally refers to Rn-222. Another isotope, Rn-220 (known as 'thoron'), is given off by thorium, which is a constituent of many Australian mineral sands. See also Appendix 2.

[3] mSv/yr = millisievert per year, a measure of dose. See Appendix 1.

underground mine, a good forced ventilation system is required to achieve the same result. At the Olympic Dam mine (an underground mine) in Australia, radiation doses are kept very low, with an average of less than 1 mSv/yr. In Canadian underground mines, doses average about 3 mSv/yr.

- Strict hygiene standards to prevent inhalation or ingestion, similar to those in a lead smelter, are imposed on workers handling the uranium oxide concentrate. If it is ingested it has a chemical toxicity similar to that of lead oxide[4]. Respiratory protection is used in particular areas identified by air monitoring, or where there could be a hazard.

These precautions with respect to radon are a relatively new response to an old hazard. From the 15th Century, many miners who had worked underground in the mountains near the present border between Germany and the Czech Republic contracted a mysterious illness, and many died prematurely. In the 1870s the illness was diagnosed as lung cancer, but it was not until 1921 that radon gas was suggested as the possible cause. Although this was confirmed by 1939, between 1946 and 1959 a lot of underground uranium mining took place in the USA without the precautions which might have become established as a result of the European experience. In the early 1960s, a higher than expected incidence of lung cancer became evident among these miners who smoked. The cause was then recognized as the emission of alpha particles from radon and, more importantly, its solid daughter products of radioactive decay. The miners concerned had been exposed to high levels of radon 10-15 years earlier, accumulating radiation doses well in excess of present recommended levels.

The small, unventilated uranium 'gouging' operations in the USA which led to the greatest health risk are a thing of the past. In the last 40 years, individual mining operations have been larger, and efficient ventilation and other precautions routinely protect underground miners from these hazards. Open cut mining of uranium virtually eliminates the danger. There has been no known case of illness caused by radiation among uranium miners in Australia or Canada, and it is clear that no major occupational health effects have been experienced in either country.

After mining is complete, most of the orebody, with virtually all of the radioactive radium, thorium and actinium materials, will end up in the tailings dam[5]. Hence radiation levels and radon emissions from tailings may be significant, comparable with those from outcropping orebodies. Therefore, the tailings need to be covered over with enough rock, clay and soil to reduce both gamma radiation levels and radon emanation rates to levels near those naturally occurring in the region. A vegetation cover can then be established. Any regional increase in radon release due to mining operations is very small and not measurable outside the mine lease.

Process water from which tailings solids have settled out contains radium and other metals that would be undesirable in the outside environment. This water is never released to natural waterways, but is stored in tailings retention areas and evaporated or treated for re-use. Either way, the contained metals are kept in safe storage, as in an orebody.

At Ranger in tropical north Australia, rainfall run-off is segregated in accordance with water quality, and high quality water (where radionuclide levels do not exceed drinking water standards) from relatively undisturbed catchments is released during flood times. Poorer quality water is retained on site and treated.

5.2 THE NUCLEAR FUEL CYCLE

Fuel cycles describe the way in which fuel gets to where it is used to provide energy and what happens to it and its wastes afterwards. The 'front end' of the nuclear fuel cycle covers all the stages from uranium mining to burning of the fuel in the reactor. The 'back end' refers to all stages subsequent to removal of used fuel from the reactor.

All aspects of obtaining and preparing the uranium fuel, using it, and the management of used fuel together make up what is known as the nuclear fuel cycle. As the term suggests it was originally the intention with nuclear power to recycle the unused part of the spent fuel so that it is incorporated into the fresh fuel elements. However, more

---

4 Both lead and uranium are toxic and affect the kidney. The body progressively eliminates most Pb or U via urine.

5 About 95% of the radioactivity in the ore is from the U-238 decay series (see Appendix 2), totalling about 450 kBq/kg in ore with 0.3% U_3O_8. The U-238 series has 14 radioactive isotopes in secular equilibrium, thus each represents about 32 kBq/kg (irrespective of the mass proportion). When the ore is processed, the U-238 and the very much smaller masses of U-234 (and U-235) are removed. The balance becomes tailings, and at this point has about 85% of its original intrinsic radioactivity. However, with the removal of most U-238, the following two short-lived decay products in the uranium decay series (Th-234 and Pa-234) soon disappear, leaving the tailings with a little over 70% of the radioactivity of the original ore after several months. The controlling long-lived isotope then becomes Th-230 which decays with a half-life of 77,000 years to radium-226 followed by radon-222. (Supervising Scientist Group, Australia).

commonly this is not done today because fresh supplies of uranium are relatively inexpensive.

Unlike coal, uranium as mined cannot be fed directly into a power station. It has to be purified, isotopically concentrated (usually) and made up into special fuel rods. Figure 16 shows the so-called 'open fuel cycle' for nuclear power, which is the system as it stands today in most countries using the most common kinds of reactors.

Starting in uranium mines, ore is mined and milled to produce uranium in the form of uranium oxide concentrate, essentially U_3O_8, as described in Section 5.1. This material, a khaki-coloured powder after drying, is sold to customers and shipped from the mine. It has the same isotopic ratio as the ore, where uranium-235 (U-235) is present to the extent of about 0.7%. Apart from traces of U-234, the rest is a heavier isotope of uranium – U-238. Most reactors, including the common light water types, cannot run on this natural uranium, so the proportion of U-235 must be increased to between 3% and 5%. This process is called enrichment.

Enrichment is a fairly high-technology physical process which requires the uranium to be in the form of a gas. The simplest way to achieve this is to convert the uranium oxide to uranium hexafluoride (UF_6, often referred to as 'hex'), which is a gas at little more than room temperature (actually at 56°C). Hence the first destination of uranium oxide concentrate from a mine is a conversion plant where it is purified and converted to UF_6.

The UF_6 is then fed through an enrichment plant[6] which increases the proportion of the fissile U-235 isotope about five- to seven-fold from the 0.7% of U-235 found in natural uranium. In this physical process about 85% of the natural uranium feed is rejected as 'depleted uranium'

Figure 16. The open nuclear fuel cycle

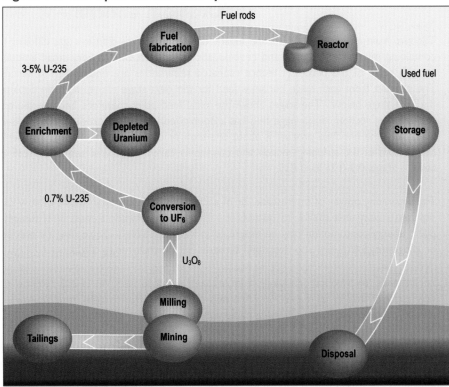

or 'tails' (mainly U-238) which is stockpiled[7]. Thus, after enrichment about 15% of the original quantity is available as enriched uranium containing 3-5% U-235.

The enrichment methods now in use are based on the slight difference in atomic mass of U-235 and U-238. Most enrichment today is with high-speed centrifuges, which draw off the lighter fraction with U-235 from the centre, and the heavier molecules with U-238 from the outside. An older process, gaseous diffusion, passed the UF_6 gas through a long series of membrane barriers which allow the lighter molecules with U-235 through faster than the U-238 ones.

Enriched uranium then goes on to a fuel fabrication plant where the reactor fuel elements are made. The UF_6 is converted to uranium dioxide (UO_2), which is formed into small cylindrical pellets about 2 cm long and 1.5 cm in diameter. These are heated strongly to become hard ceramic pellets, and then loaded into zirconium alloy or stainless steel tubes about 4 metres long to form fuel rods. These are assembled into bundles about 300 mm square to form reactor fuel assemblies. Fuel assemblies

---

[6] Enrichment was originally undertaken using the expensive and energy-intensive gaseous diffusion process. Newer plants are based on very much more efficient gas centrifuge technology. The next generation of enrichment plants may use advanced laser technology.

[7] This material cannot be used in current types of reactors; its only significant use is as a feed for fast neutron reactors, or to dilute ex-military uranium – see Sections 4.6 & 3.4. It is stored as UF_6 in steel cylinders as a liquid or solid. Usually less than 0.3% U-235 remains in it.

Uranium enrichment

The two main enrichment (or isotope separation) processes are diffusion (gas diffusing under pressure through a membrane containing microscopic pores) and centrifugation. In each case, a very small amount of isotope separation takes place at each stage of the process. Hence repeated separations are undertaken in successive stages, arranged in a cascade. The product from each stage becomes feed for the next stage above, and the depleted material is added to the feed for the next stage below. The stages above the initial feed point therefore become the enriching section and those below are the stripping section. Each stage thus has a double feed (enriched product from below and depleted product from above). Ultimately, the enriched product is about one-sixth or one-seventh the amount of depleted material, which is drawn off at the bottom of the stripping section and commonly called 'tails'. The residual U-235 concentration in the tails is the tails assay – about 0.2-0.25% U-235.

The separating power of the cascade, or of each stage, is described in terms of flow capacity and enriching ability, using the separative work unit (SWU) to quantify it. This is dimensionally a mass unit, though it indicates energy (for a particular plant, energy consumption may be described in kWh per SWU). Since feed or product quantities are measured in tonnes or kilograms, SWUs are also described similarly, but normally as kg SWU.

For instance, to produce one kilogram of uranium enriched to 3.5% U-235 requires 4.8 SWU if the plant is operated at a tails assay of 0.25%, or 5.5 SWU if the tails assay is 0.20% (thereby requiring only 6.5 kg instead of 7.1 kg of natural U feed). Here, and in the following paragraph, kg SWU units are implied.

About 100,000 to 120,000 SWU is required to enrich the annual fuel loading for a typical 1000 MWe light water reactor. Enrichment costs are related to electrical energy used. The gaseous diffusion process consumes up to 2,400 kWh (8,600 MJ) per SWU, while gas centrifuge plants require only about 50 kWh per SWU (180 MJ).

Diffusion process

The old diffusion process relied on a difference in average velocity of the two types of UF6 molecules to drive the lighter ones more readily through holes in the membranes, since they move faster. Each stage consists of a compressor, a diffuser and a heat exchanger to remove the heat of compression. The enriched UF6 product is withdrawn from one end of the cascade and the depleted UF_6 is removed at the other end. The gas must be processed through some 1400 stages to obtain a product with a concentration of 3% to 4% U-235. Diffusion plants typically have a small amount of separation through one stage (hence the large number of stages) but are capable of handling large volumes of gas.

Centrifuge process

Centrifuge enrichment relies on the simple mass difference of the molecules coupled with the peripheral velocity in a rapidly rotating cylinder (the centrifuge rotor). Countercurrent movement of gas within the rotor due to a thermal gradient enhances this effect. The gas is fed into a series of evacuated cylinders, each containing a rotor about 3-5 metres long and 20 cm diameter. When the rotors are spun rapidly, the heavier molecules with U-238 increase in concentration towards the cylinder's outer edge, leaving a corresponding increase in concentration of molecules with U-235 near the centre. The countercurrent flow enables enriched product to be drawn off axially, heavier molecules at one end and lighter ones at the other.

To obtain efficient separation of the two isotopes, centrifuges rotate at very high speeds, typically 50,000 to 70,000 rpm, with the outer wall of the spinning cylinder moving at between 400 and 500 metres per second, to give a million times the acceleration of gravity. There are considerable materials and engineering challenges in producing such equipment, and carbon fibre is the main rotor material.

Centrifuges at Urenco enrichment plant (Urenco)

Although the volume capacity of a single centrifuge is much smaller than that of a single diffusion unit, its ability to separate isotopes is much greater. Centrifuge stages normally consist of a large number of centrifuges in parallel. Such stages are then arranged in cascade similarly to those for diffusion. In the centrifuge process, the number of stages may only be 10 to 20, instead of the thousand or more required for diffusion.

Laser process

Laser isotope separation processes have been a focus of interest for some time. They promise lower energy inputs, lower capital costs and lower tails assays, hence significant economic advantages. Laser separation processes may in principle use either atomic or molecular gases, but atomic processes have failed to work. A molecular laser separation process is currently under intensive development and there are plans to build the first full-scale Global Laser Enrichment plant in the USA, based on Australian technology.

Laser processes utilise the very precise beam frequencies characteristic of lasers. The interaction of the laser beam with a gas enables it to exploit the excitation or ionisation of isotope-specific atoms in the vapour. The process nearing commercialisation works on the principle of photo-dissociation of UF_6 to UF_5^+, using tuned lasers to break the molecular bond holding one of the six fluorine atoms to a U-235 atom. This then enables the ionised UF_5^+ to be separated from the unaffected UF_6 molecules containing U-238 atoms, hence achieving a separation of isotopes.

A fuller description of enrichment processes is in the WNA information paper on *Uranium Enrichment*.

of this kind are used to power the common light water power reactors, both BWRs and PWRs (see Table 8 on page 42). A 1000 MWe reactor has about 75 tonnes of fuel in it, in about 170 fuel assemblies (if PWR).

Canadian CANDU (CANadian Deuterium Uranium) reactors have a different design, and run on natural (*i.e.* unenriched) uranium. Instead of a single large pressure vessel containing the core, they have multiple (*e.g.* 300-600) horizontal pressure tubes, each containing fuel and heavy water coolant. The pressure tubes extend through the reactor vessel, or calandria, which contains the heavy water moderator[8]. CANDU fuel bundles are only 10 cm diameter and 50 cm long.

Inside all kinds of operating reactors a fission chain reaction occurs in the fuel rods, as described in Sections 3.1 and 3.7. Fast neutrons are slowed by the water, heavy water or graphite moderator so that they can cause fission. Neutron-absorbing control rods are inserted or withdrawn to regulate the speed of the reaction, though normally reactors are run at full power. Heat from the fission reaction is conveyed from the reactor core by the coolant and is used to make steam, which in turn is used to generate electricity.

In a light water reactor the fuel stays in the reactor for about three to four years, generating heat from fission of

both the U-235 and also the fissile plutonium (*e.g.* Pu-239), which is formed there from U-238. Progressively over a few years, the level of fission products and other neutron-absorbers builds up so that they interfere with the fission chain reaction, and the used fuel assemblies are therefore removed. About one-third of the fuel may be changed each year to 18 months.

When removed, used fuel is hot and radioactive. It is therefore stored under water to remove the heat and to provide shielding from radiation, pending the next step. The used fuel may be reprocessed in the case of countries such as UK, France and Japan, which have chosen to 'close' the fuel cycle, or the next step may be final disposal in the case of countries such as the USA, Canada and Sweden, which have chosen the 'open fuel cycle'. Storage is initially at the reactor site. After a couple of years it may be transferred to dry storage casks on site, or moved elsewhere.

The closed fuel cycle is illustrated in the more complex diagram in Figure 17. In Europe, reprocessing operations have been going on for many years to recycle the useful content. Today, there is greater interest in it in the USA.

In the closed fuel cycle, fuel is supplied in exactly the same way as before. Starting with uranium mines and mills, the uranium goes through conversion, enrichment,

---

[8] Heavy water, or deuterium oxide (sometimes: D_2O), contains deuterium, which is an isotope of hydrogen having a neutron in the nucleus.

57

Figure 17. The closed nuclear fuel cycle

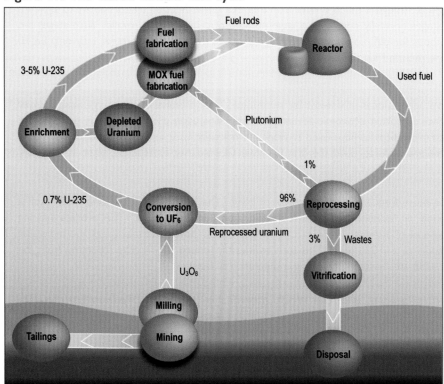

very good nuclear fuel which needs no enrichment process. It can be mixed with depleted uranium, made into fuel rods in a mixed oxide (MOX) fuel fabrication plant, and put back into the reactor as fresh fuel (see Section 6.2). Alternatively, it could be used to fuel future fast neutron reactors (see Section 4.6 and Figure 18).

The recovered uranium can go back to be enriched (via conversion), and on into fresh fuel for a reactor, though at present most is simply stockpiled. The closed fuel cycle is thus a more efficient system for making maximum use of the uranium dug out of the ground (by about 30%, in energy terms) and that is why the industry originally favoured this approach. However, due largely to many years of low uranium prices (mid-1980s to about 2003) and concerns about separating plutonium, plans for widespread reprocessing of used reactor fuel have not eventuated. Several European countries, Russia, China and Japan are proceeding with the closed fuel cycle, and across Europe over 35 reactors are licensed to load 20-50% of their core with MOX fuel containing up to 7% reactor-grade plutonium.

and fuel fabrication to the reactor. But after being removed from the reactor and stored for a short time, the used fuel rods are put through a reprocessing plant where they are chopped up and dissolved in acid. Various chemical processes recover and separate the two valuable components: plutonium and unused uranium. This leaves about 3% of the fuel as separated high-level waste. After drying, it is reduced to a small volume and is vitrified, resulting in a solid highly radioactive material suitable for permanent disposal (see also Sections 6.2 and 6.4).

About 96% of the uranium which goes into the reactor emerges again in the used fuel, albeit depleted to less than 1% U-235. Some of what has been used up was converted into heat and radioactive fission products and some into plutonium and other actinide elements (see Figure 19 on page 61). Hence, reprocessing used fuel has some economic benefits in recovering the unused uranium, along with the plutonium that has been generated and not burned in the reactor. It also substantially reduces the volume of material to be disposed of as high-level waste, which has further economic benefit.

Plutonium (which is reactor-grade[9]) comprises about 1% of the used fuel. It is a mixture of isotopes and makes a

Cylinders of depleted uranium hexafluoride in storage (US Department of Energy)

---

[9] Reactor-grade plutonium has about one-third non-fissile isotopes and is thus very different from weapons-grade material.

In the closed fuel cycle (see Figure 17), it can be seen that conventional reactors give rise to three 'surplus' materials: depleted uranium (from enrichment), plutonium (from neutron capture in the reactor core, separated in reprocessing), and recycled uranium from reprocessing (with around 1% U-235). Fast neutron reactors are able to use all of these. When that fuel cycle is set up and operating, refuelling involves adding only a little natural or depleted uranium.

Operation of the fast neutron reactor is described in Section 4.6.

5.3 THORIUM CYCLE

Near-breeder or thorium cycle reactors are similar to fast breeders in that a fertile material – naturally-occurring thorium (Th-232) – will absorb slow neutrons to become (indirectly) uranium-233 (U-233) which, like U-235, is fissile. Given a start with some fissile material (U-233, U-235 or Pu-239) as a driver, a breeding cycle similar to but more efficient than that with U-238 and plutonium (in conventional reactors) can be set up. The chain reaction yields heat, while surplus neutrons convert more thorium to U-233.

The technology is considered by some to be attractive because plutonium (and other long-lived transuranic elements) production is avoided, abundant thorium is used as a fuel, and the efficiency of fuel use approaches

Figure 18. The fast neutron reactor fuel cycle

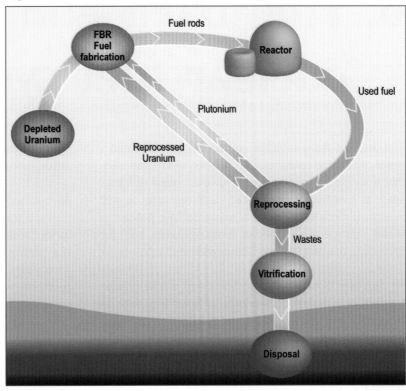

that of the fast breeder reactor. In operation, the amount of fissile uranium produced is not quite enough to sustain the reaction, hence the term 'near-breeder' is generally used, and some input of fissile U-235 or Pu-239 will always be required. Also it is necessary to reprocess the used fuel to recover the U-233 and recycle it, but such reprocessing of thorium fuel has not yet been done on any scale.

Though a focus of interest for 40 years, only India is developing the concept via a three-stage program which will culminate in the advanced heavy water reactors (briefly described in Section 4.2). An alternative reactor concept for thorium is the liquid fluoride thorium reactor (LFTR), a type of molten salt reactor, and China is pursuing R&D on this.

There is also a joint Canadian-Chinese project investigating the use of thorium in CANDU reactors.

See also WNA information paper on *Thorium*.

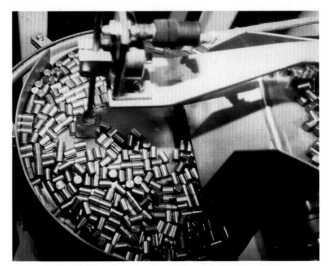

MOX fuel pellets at the Melox fabrication plant, Marcoule, France (Photograph: IAEA)

6. The 'back end' of the nuclear fuel cycle

The used fuel from a nuclear reactor can be considered a waste, and some countries have policies which determine that it should be treated thus. However, many countries see it as containing significant recyclable resources as well as what is more strictly seen as waste, so have policies to enable reprocessing to recover those resources, both for immediate use and for longer-term reserves.

As discussed briefly in Section 5.2, there is now greater interest in recycling components of used fuel both as resources, and also in order to burn long-lived actinides, making high-level wastes much less long-lived radiologically, and hence making disposal easier. This is reversing a widespread sentiment set in train by a US policy decision in 1977.

6.1 NUCLEAR WASTES

Despite its demonstrable safety record over more than half a century, one of the most controversial aspects of the nuclear fuel cycle today is the question of management and disposal of radioactive wastes.

The most difficult of these are the high-level wastes, which have nevertheless been handled, transported and stored for many decades virtually without incident, and certainly without harm to anyone. There are two alternative strategies for managing such wastes:

- Reprocessing used fuel to separate the wastes from the recycled fuel materials (followed by vitrification and disposal of those actual wastes).
- Direct disposal of the used fuel, which contains high levels of radioactivity, as waste.

The principal nuclear wastes, which have no further use in any scenario and are only a small proportion of the used fuel, remain locked up securely in the ceramic reactor fuel (direct disposal) or in borosilicate glass (after reprocessing and vitrification).

As outlined in Chapters 3 and 5, 'burning' the fuel of the reactor core produces fission products such as various isotopes of barium (Ba), strontium (Sr), caesium (Cs), iodine (I), krypton (Kr) and xenon (Xe) – see Figure 15 on page 38. Many of the isotopes formed as fission products within the fuel are highly radioactive, and correspondingly short-lived.

As well as these smaller atoms formed from the fissile portion of the fuel, various transuranic isotopes or actinides are formed by neutron capture (see "Neutron capture: transuranic elements and activation products" in Section 3.7). These include plutonium-239 (Pu-239), Pu-240 and Pu-241[1], as well as others arising from some of the U-238 in the reactor core by neutron capture and subsequent beta decay. All are radioactive and apart from much of the fissile plutonium which is 'burned', they remain within the used fuel when it is removed from the reactor. The transuranic isotopes and other actinides[2] form most of the long-lived portion of high-level wastes.

Today there is renewed interest in reprocessing used fuel to recover both uranium and plutonium for recycling, and also to recover the long-lived transuranics so that the remaining high-level waste is more easily disposed of due to its radioactivity being shorter-lived. This interest is coupled with the prospect of much greater use of fast reactors after the mid-2020s, which are best able to fission such actinides[3].

While the civil nuclear fuel cycle generates various wastes, many of them potentially hazardous, these do not become pollution since virtually all are contained and managed. In fact, nuclear power is the only major energy-producing industry which takes full responsibility for all its wastes and fully costs this into the product. Furthermore, the

---

[1] It is Pu-241 which decays to give americium-241, which is used in household smoke detectors.
[2] Actinides are elements with atomic number of 89 (actinium) or above; transuranics are those above 92 (uranium).
[3] The fast neutron environment minimises neutron capture reactions and maximises fissions in actinides.

expertise developed in managing civil wastes is now starting to be applied to military wastes, which pose a real environmental problem or hazard in a few parts of the world.

Radioactive wastes are normally classified as low-level, intermediate-level or high-level wastes, according to the amount and types of radioactivity in them. In a country such as the UK, radioactive wastes comprise about 1% of all toxic wastes which require management and disposal.

Another factor in managing radioactive wastes is the time that they are likely to remain hazardous. This depends on the kinds of radioactive isotopes in them, and particularly the half-lives characteristic of each of those isotopes. The half-life is the time it takes for a given amount of a radioactive isotope to lose half of its radioactivity. After four half-lives the level of radioactivity is 1/16 of the original and after eight half-lives 1/256.

> **Radioactive wastes comprise a variety of materials requiring management to protect people and the environment. Management and disposal is technically straightforward.**

The various radioactive isotopes arising from nuclear energy have half-lives ranging from fractions of a second to minutes, hours or days, through to billions of years. Radioactivity decreases with time as these isotopes decay into stable, non-radioactive ones. The rate of decay of an isotope is inversely proportional to its half-life; a short half-life means that it decays rapidly. Hence, for each kind of radiation, the higher the intensity of radioactivity

in a given amount of material, the shorter the half-life (or half-lives) involved.

Three general principles are employed in the management of radioactive wastes:

- Concentrate and contain.
- Dilute and disperse.
- Delay and decay.

The first two are also used in the management of non-radioactive wastes. The wastes are either concentrated and then isolated permanently, or small quantities are diluted to acceptable levels (often with a delay to allow decay) and then discharged to the environment. 'Delay and decay' however is unique to radioactive waste management; it means that the waste is stored and its radioactivity is allowed to decrease naturally through decay of the radioisotopes in it. When the radioactivity has decreased to normal background levels it can be dispersed.

High-level wastes

In the civil nuclear fuel cycle the main focus of attention is high-level waste containing the fission products and transuranic elements formed in the reactor core.

The high-level waste may be used fuel itself, or the principal waste arising from reprocessing this fuel. Either way, the volume is modest – about 25-30 tonnes of used fuel, or three cubic metres of vitrified waste, per year for a typical large nuclear reactor (1000 MWe, light water type). This can be effectively and economically isolated. Its level of radioactivity falls rapidly (see Figure 21 on page 68, though note that the curve shown starts at one year – the fall is more rapid

Figure 19. What happens in a light water reactor over three years

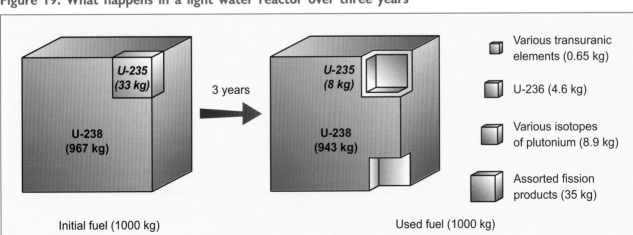

Source: Scientific American, June 1977

Final repository for intermediate-level wastes – SFR, Sweden (SKB)

before that). For instance, a newly-discharged light water reactor fuel assembly is so radioactive that it emits several hundred kilowatts of heat, but after a year this is down to 5 kilowatts and after five years, to one kilowatt. In 40 years the radioactivity in it drops to about $^1/_{1000}$ of the level at discharge.

If the used fuel is reprocessed, about 3% of it emerges as liquid high-level waste, containing the 'ash' from 'burning' uranium. This separated waste consists of the highly-radioactive fission products and some heavy elements with long-lived radioactivity. It generates a considerable amount of heat and requires cooling. This is dried and vitrified into borosilicate glass (similar to Pyrex) for encapsulation, interim storage, and eventual disposal deep underground. This is the policy adopted by the UK, France, Germany, Switzerland, Japan, China and India (see Sections 6.2 & 6.4).

On the other hand, if used reactor fuel is not reprocessed, all the highly radioactive fission product isotopes and the much smaller quantity of long-lived actinides remain in it, and so whole fuel assemblies are treated as high-level waste. The direct disposal option is being pursued by the USA, Canada, Finland and Sweden (see Section 6.4)[4].

High-level wastes make up only 2-3% of the volume of all radioactive wastes worldwide, but they hold 95% of the total radioactivity in them.

Low-level wastes

In addition to the high-level wastes from nuclear power production, all use of radioactive materials in hospitals, laboratories, universities and industry generates what are termed low-level wastes (cleaning equipment, gloves, clothing, tools, *etc.*), which are not dangerous to handle but must be disposed of more carefully than normal garbage. They may be incinerated to reduce their volume. Ultimately, they are usually buried in shallow landfill sites. Provided all highly toxic materials are first separated and included with intermediate-level wastes (see below), this has been shown to be an effective means of waste management for such relatively innocuous materials. Many countries have final repositories in operation for low-level wastes, which have about the same level of radioactivity as a low-grade uranium orebody. They amount to some 50 times the volume of the annual arisings of high-level wastes. Worldwide they make up 90% of the volume but have only 1% of the total radioactivity of all radioactive wastes.

Some low-level liquid wastes from reprocessing plants are discharged to the sea. These include radionuclides which are distinctive, notably technetium-99 (sometimes used as a tracer in environmental studies), and this can be discerned many hundred kilometres away. However, such discharges are regulated and controlled, and do not add to natural background levels of radiation.

Nuclear power stations and reprocessing plants release small quantities of radioactive gases (krypton-85 and xenon-133) and trace amounts of iodine-131 to the atmosphere. However, they have short half-lives, and the radioactivity in the emissions is diminished to a safe level by delaying their release. Also krypton and xenon are chemically and biologically inert. The net effect is too small to warrant consideration in any life-cycle analysis.

Intermediate-level wastes

These mostly come from the nuclear industry and research reactors. They are more radioactive and need to be shielded from people before treatment and disposal. They typically comprise resins, chemical sludges and reactor components, as well as contaminated materials from reactor decommissioning. Mostly these wastes are embedded in concrete for disposal. Generally short-lived intermediate-level waste (mainly from reactors) is buried, but long-lived waste (some of that from reprocessing nuclear fuel) will be disposed of deep underground with high-level wastes. Worldwide it makes up about 7% of the volume of radioactive wastes and has 4% of the radioactivity.

---

[4] A number of countries have deferred choosing between reprocessing and direct disposal, and the USA is reconsidering its policy.

6.2 REPROCESSING USED FUEL

The principal reason for reprocessing is to recover unused uranium and plutonium for recycling. A secondary reason is to reduce the volume and long-term radioactivity of material to be disposed of as high-level waste.

Reprocessing used fuel avoids the waste of a valuable resource because most of the used fuel (uranium with less than 1% U-235 and a little plutonium) can be recycled as fresh fuel elements, saving some 30% of the natural uranium otherwise required. The plutonium becomes mixed oxide (MOX) fuel, and is a significant resource. The uranium can be re-enriched and recycled, though most today is simply stockpiled. The separated radioactive high-level wastes are then vitrified so that they are in a compact, stable, insoluble and solid form, which is easier to dispose of than the more bulky used fuel assemblies.

Future development of reprocessing is likely to separate the long-lived transuranic elements (plutonium together with minor actinides) to be burned in a fast reactor, leaving only shorter-lived fission products as the waste, and further simplifying disposal. A further development would be then to separate some, especially longer-lived, fission products for transmutation[5].

A 1000 MWe light water reactor produces about 27 tonnes of used fuel per year. So far, some 100,000 tonnes of used fuel from commercial power reactors has been reprocessed, and annual reprocessing capacity is nearly 5000 tonnes per year.

Table 12. World commercial reprocessing capacity

		(Tonnes per year)
Light water reactor fuel	France, La Hague	1,700
	UK, Sellafield (THORP)	900
	Russia, Ozersk (Mayak)	400
	Japan (Rokkasho)	40*
LWR total (approx)		**3,040**
Other nuclear fuels	UK, Sellafield (Magnox)	1,500
	India (PHWR)	330
Non-LWR total (approx)		**1,830**
Total civil capacity (approx)		**4,870**

* 800 t/yr expected from late 2012

Used fuel assemblies removed from a reactor are very radioactive, and produce heat. They are therefore put into large tanks or 'ponds' of water for cooling, while three metres of water over them shields the radiation. Here they remain, mostly at the reactor site, for a number of years while the level of radioactivity decreases considerably. For most types of fuel, reprocessing occurs about five years after reactor discharge. If it is long delayed, the quality of the plutonium degenerates due to Pu-241 (about 12% of Pu, and fissile) decaying into americium.

Areva's La Hague reprocessing plant on the Cotentin peninsula, Normandy, France

---

[5] Notably iodine-129, technetium-99, caesium-135 and strontium-90.

Thermal Oxide Reprocessing Plant at the UK's Sellafield site
(Sellafield Ltd)

Used fuel may be transported after initial cooling, using special shielded casks which hold only a few (e.g. six) tonnes of used fuel but weigh about 100 tonnes – see Panel on *Transporting radioactive materials*. Transport of used fuel and other high-level waste is tightly regulated.

Reprocessing of used oxide fuel involves dissolving the fuel elements in nitric acid. Chemical separation of uranium and plutonium is then undertaken. The Pu and U can be returned to the input side of the fuel cycle – the plutonium straight to fuel fabrication and the uranium to the conversion plant prior to re-enrichment, though in fact most is put into long-term storage. Although Figure 17 on page 58 shows reprocessing and MOX fuel fabrication on opposite sides of the diagram, for recycled fuel these facilities may be located on a single site. The remaining material after Pu and U are removed is high-level waste, comprising about 3% of the used fuel. It is highly radioactive and continues to generate a lot of heat.

After reprocessing, the recovered uranium requires re-enrichment so if it is immediately recycled it goes first to a conversion plant. This is complicated by the presence of impurities and two new isotopes in particular, U-232 and U-236, which are formed by neutron capture in the reactor. Both decay much more rapidly than U-235 and U-238, and one of the daughter products of U-232 emits very strong gamma radiation, which means that shielding is necessary in the enrichment plant. U-236 is a neutron absorber, which impedes the chain reaction and means that a higher level of enrichment is required in the recycled uranium to compensate. Being lighter, both isotopes tend to concentrate in the enriched (rather than depleted) output, so reprocessed uranium which is re-enriched for fuel must be segregated from enriched fresh uranium.

A great deal of reprocessing has been going on since the 1940s, initially for military purposes, to recover plutonium (from low burn-up fuel used in special reactors) for weapons. In the UK, metal fuel elements from the first generation gas-cooled commercial reactors have been reprocessed at Sellafield for about 45 years. The 1500 t/yr Magnox fuel reprocessing plant has been successfully developed to keep abreast of evolving safety, hygiene and other regulatory standards. From 1969 to 1973 oxide fuels were also reprocessed, using part of the plant modified for the purpose. The Thermal Oxide Reprocessing Plant (THORP) was commissioned in 1994 but has never run at full capacity.

In the USA, three plants for the reprocessing of civilian oxide fuels were built, but for technical, economic and political reasons all were shut down or aborted. In all, the USA has over 250 plant-years of reprocessing operational experience, the vast majority being at government-operated defence plants since the 1940s.

In France, one 400 t/yr reprocessing plant operated for metal fuels from early gas-cooled reactors at Marcoule. At La Hague, reprocessing of oxide fuels has been carried out since 1976, and two 800 t/yr plants are now operating reliably. India has a 100 t/yr oxide fuel plant operating at Tarapur with others at Kalpakkam and Trombay, and Japan is commissioning a major 800 t/yr plant at Rokkasho, having had most of its used fuel reprocessed in Europe meanwhile. It had a small (100 t/yr) plant operating to 2006. Russia has a 400 t/yr oxide fuel reprocessing plant at Ozersk.

It is worth noting that wastes from weapons programs will continue to overshadow civil nuclear wastes in countries like the USA and Russia for some decades, no matter how rapidly commercial nuclear power expands. The legacy of these wastes, dating from the 1940s, in polluted land and leaking storage tanks – and the prospect of very high clean-up costs – remains with those countries which produced them.

Mixed oxide fuel

Separated plutonium is recycled via a dedicated mixed oxide (MOX) fuel fabrication plant. In France the reprocessing output is coordinated with MOX plant input, to avoid building up stocks of plutonium. (If separated plutonium is stored for some years the level of americium-241 – the isotope used in household smoke detectors – will accumulate and make it difficult to handle through a MOX plant due to the elevated levels of gamma radioactivity.)

Table 13. World mixed oxide fuel fabrication capacities

	Tonnes per year	
	2009	2015
France, Melox	195	195
Japan, Tokai	10	10
Japan, Rokkasho	0	130
Russia, Mayak, Ozersk	5	5
Russia, Zheleznogorsk	0	60
UK, Sellafield	40	0
Total for LWR	**250**	**400**

See also WNA information paper *Processing of Used Nuclear Fuel*, which includes new reprocessing technologies, and WNA information paper *Mixed Oxide Fuel*.

Used fuel cask being loaded on to ship

6.3 HIGH-LEVEL WASTES FROM REPROCESSING

Though the quantities involved are small (see Section 6.1), high-level wastes from reprocessing used reactor fuel require great care in handling, storage and disposal because they are very radioactive and therefore hot. They comprise fission products and transuranic elements which emit alpha, beta and gamma radiation at high levels, as well as a lot of heat. The heat arises mainly from the fission products, which mostly have the shorter half-lives. These are the materials popularly thought of as 'nuclear wastes'. In future developments of reprocessing, these high-level wastes will comprise mainly fission products, which means that in a 100-year perspective they are little different from today's separated wastes, but taking a 1000-year perspective, they are much less radioactive.

Transporting radioactive materials

Nuclear materials have been transported since before the advent of nuclear power some 60 years ago. Many of these are similar to materials used in other industrial activities. However, the nuclear industry's fuel and waste materials are radioactive, and it is these 'nuclear materials' which are the main focus of regulation. The procedures employed are designed to ensure the protection of the public and the environment.

About 20 million shipments of radioactive material (which may be either a single package or a number of packages sent from one location to another at the same time) take place around the world each year. Radioactive material is not unique to the nuclear fuel cycle and most shipments of such material are not fuel cycle related. Radioactive materials are used extensively in medicine, agriculture, research, manufacturing, non-destructive testing and minerals exploration.

However, transport is an integral part of the nuclear fuel cycle. There are some 435 nuclear power reactors operable in 30 countries but uranium mining is viable in only a few areas. Furthermore, a limited number of specialised facilities have been developed in various locations around the world to provide fuel cycle services. It is clear that there is a need to transport nuclear fuel cycle materials to and from these facilities. Indeed, most of the material used in nuclear fuel is transported several times on its way through the fuel cycle. Transport operations are frequently international, and are often over large distances. Nuclear materials are generally transported by specialised transport companies.

Since 1971 there have been some 7000 shipments of used fuel (over 35,000 tonnes) over more than 30 million kilometres with no property damage or personal injury, no breach of containment, and very low dose rate to the personnel involved (e.g. 0.33 mSv/yr per operator at La Hague). Some 300 sea voyages have been made carrying used nuclear fuel or separated high-level waste over a distance of more than 8 million kilometres. The major company involved has transported over 4000 casks, each of about 100 tonnes, carrying 8000 tonnes of used fuel or waste. A quarter of these have been through the Panama Canal.

In Sweden, more than 80 large transport casks are shipped annually from nuclear power stations (all on the coast) to the Central Interim Storage Facility for Spent Nuclear Fuel (CLAB). A purpose-built 2000 tonne ship is used for moving the used fuel. In the USA, more than 3000 shipments of used nuclear fuel have been made over 2.7 million kilometres with no harmful release of radiation.

Packaging

The principal assurance of safety in the transport of nuclear materials is the design of the packaging, which must allow for foreseeable accidents. The consignor bears primary responsibility for this.

'Type A' packages are designed to withstand minor accidents and are used for medium-activity materials such as medical or industrial radioisotopes. Ordinary industrial containers are used for low-activity material such as U_3O_8.

Packages for high-level waste (HLW) and used fuel are robust and very secure containers known as 'Type B' packages. They also maintain shielding from gamma and neutron radiation, even under extreme conditions. There are over 150 kinds of Type B packages, and the larger ones weigh up to 110 tonnes each when empty and hold up to 6 tonnes of used fuel.

In France alone, there are some 750 shipments each year of Type B packages, among 15 million shipments of goods classified as 'dangerous materials', 300,000 of these being radioactive materials of some kind.

Smaller amounts of high-activity materials (including plutonium) transported by aircraft will be in 'Type C' packages, which give greater protection in all respects than Type B packages in accident scenarios.

To limit the risk in the handling of used nuclear fuel and vitrified HLW, dual-purpose containers (casks), which are appropriate for both storage and transport, are often used.

Regulation

Since 1961 the International Atomic Energy Agency (IAEA) has published advisory regulations for the safe transport of radioactive material. These regulations have come to be recognised throughout the world as the uniform basis for both national and international transport safety requirements in this area. Requirements based on the IAEA regulations have been adopted in about 60 countries, as well as by the International Civil Aviation Organization (ICAO), the International Maritime Organization (IMO), and regional transport organisations.

The fundamental principle is that the protection comes from the design of the package, regardless of how the material is transported.

See also WNA information paper *Transport of Radioactive Materials*.

The liquid wastes generated in reprocessing plants are stored temporarily in cooled multiple-walled stainless steel tanks surrounded by reinforced concrete. These liquids need to be changed into compact, chemically inert solids before considering the question of permanent disposal. The main method of solidifying liquid wastes is vitrification. The Australian Synroc (synthetic rock) is a more sophisticated way to immobilise such waste, but development of this is focused in specialist areas, and has moved from ceramic to composite glass-ceramic wasteforms.

Commercial vitrification plants are based on calcining of the wastes (evaporation to a dry powder), followed by incorporation in borosilicate glass. The molten glass is mixed with the dry wastes and poured into large stainless steel canisters, each holding 400kg. A lid is then welded on. A year's waste from a 1000 MWe reactor is contained in 5 tonnes of such glass, or about 12 canisters each 1.3 metres high and 0.4 metres diameter. These are stored vertically in silos, ten deep.

The 100,000 tonnes of used fuel so far reprocessed worldwide will have been reduced to about 6500 cubic metres of vitrified high-level wastes – a 20m x 20m pile 17m high in visual terms – see also Figure 20.

Vitrification processes have been developed and tested in pilot plants since the 1960s. In the UK at Harwell several tonnes of high-level wastes from reprocessed fuel were vitrified by 1966, but research was then set aside until there were enough such wastes to give the matter a higher priority. High-temperature leaching tests on this glass showed that it has remained insoluble even where some physical breakdown of the glass had occurred. Similar results have been obtained on French wastes vitrified between 1969 and 1972.

Vitrification of civil high-level radioactive wastes first took place on an industrial scale in France in 1978. It is now carried out commercially at facilities in Belgium, France and the UK with a combined capacity of 2500 canisters (1000 tonnes) per year. A new Japanese vitrification plant is being commissioned.

In 1996, two vitrification plants were opened in the USA. One, at West Valley, NY, was to treat 2.2 million litres of high-level waste from civil nuclear fuel reprocessed there 25 years earlier, the other was at Savannah River, SC, to vitrify a larger quantity of military waste.

Vitrified wastes will be stored for some time before final disposal, to allow heat and radioactivity to diminish.

Figure 20. Vitrified waste (simulated)

Borosilicate glass from the first waste vitrification plant in the UK in the 1960s. This block contains material chemically identical to high-level waste from reprocessing used fuel. A piece this size from modern vitrification plants would contain the total high-level waste arising from nuclear electricity generation for one person throughout their lifetime.

In general, the longer the material can be left before disposal the easier it is to handle and the less space is required in a repository. Depending on the actual disposal methods adopted, there will be some 50 years between reactor and disposal.

All handling of such materials involves the use of protective shielding and procedures to ensure the safety of people involved. As in all situations where gamma radiation is a feature, the simplest and cheapest protection is shielding (mass) or distance (increasing the distance by a factor of ten reduces exposure by 99%).

When separated high-level wastes (or used fuel assemblies) are moved from one place to another, robust shipping containers are used. These are designed to withstand all credible accident conditions without leakage or reduction in their radiation shielding effectiveness. Where such containers have been involved in serious accidents over the years, they have created no radioactivity hazard at all. The high standards of integrity designed into these containers also makes them difficult to breach with explosives and therefore unattractive as objects for sabotage attempts.

6.4 STORAGE AND DISPOSAL OF HIGH-LEVEL WASTES

The direct disposal option for used fuel – treating it as waste – is the policy of many countries, though usually the used fuel will be retrievable to keep options open for the future. While separated high-level wastes are vitrified to make them insoluble and physically stable, used fuel destined for direct disposal is already in a very stable ceramic form.

Used fuel storage

There are about 300,000 tonnes of used fuel in storage, much of it at reactor sites. Annual arisings of used fuel are about 12,000 tonnes, and 3000 tonnes of this goes for reprocessing. Final disposal is therefore not urgent in any logistics sense.

Sweden has had since 1988 a fully operational central long-term used fuel storage facility (CLAB) with capacity for all the country's used fuel, which is sent to it after storage at the reactor site for only a year or so. At CLAB the used fuel is handled under water, for cooling and radiological shielding, and stored for some 40 years. By 2023, this storage will be almost full and a final repository should be ready.

In considering the used fuel itself or the waste extracted from it, an important feature is the rate at which it cools and radioactivity decays. Forty years after removal from the reactor, less than one-thousandth of its initial radioactivity remains, and it is much easier to deal with (see Figure 21). This feature sets nuclear waste apart

View of CLAB interim storage facility for used fuel in Sweden (SKB)

from chemical wastes, which remain hazardous unless they are destroyed. The longer nuclear wastes are stored, the less hazardous they are, and the more readily they can be handled.

In the USA and several other countries all used fuel remains stored at reactor sites, by the utilities, and at present this is as far as the fuel cycle goes. In the USA it was intended that used fuel should be transferred from the reactor site storage ponds or dry cask storage to a federal repository at Yucca Mountain in Nevada, but this is now in doubt due to political impediments, and nuclear waste policy is under review.

International consensus on safety

Whether the final high-level waste is vitrified material from reprocessing or entire used fuel assemblies, it needs eventually to be disposed of safely. In addition to concepts of safety applied elsewhere in the nuclear fuel cycle, this means that it should not require any ongoing management after disposal, or after closure of the repository. While final disposal of civil high-level wastes will not take place for some years yet, preparations are being made in some countries. However, there is no pressing need for final disposal anywhere.

As part of an ongoing review of waste management strategies, the Radioactive Waste Management Committee of the OECD Nuclear Energy Agency reassessed the basis for geological disposal of radioactive waste from an environmental and ethical perspective. Similar consideration has been given over many years by the International Atomic Energy Agency. The question of intergenerational equity has been foremost. In a 1995 OECD report[6] and

Figure 21. Decay in radioactivity of fission products in one tonne of used PWR fuel

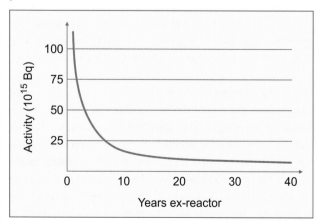

Activity (10^{15} Bq) vs Years ex-reactor

---

[6] *The Environmental and Ethical Basis of Geological Disposal of Long-Lived Radioactive Wastes*, OECD/NEA 1995 (see Appendix 3)

Figure 22. Activity of high-level waste from one tonne of used fuel

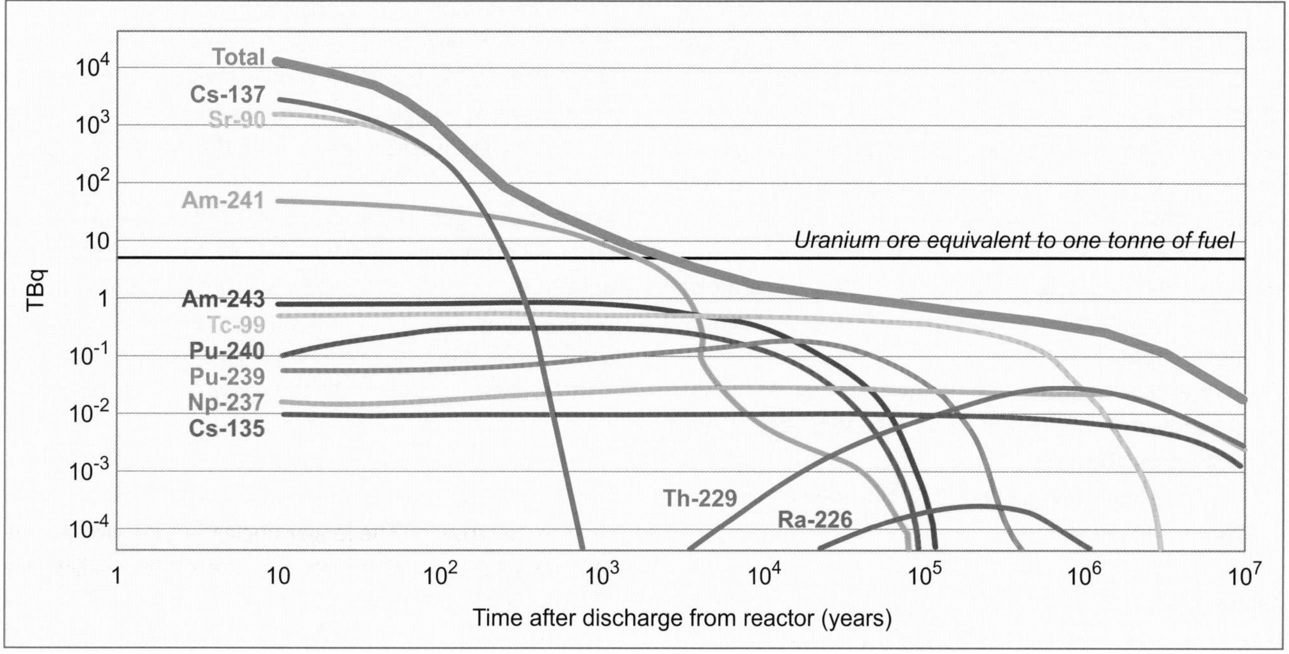

IAEA, 1992, Radioactive waste management

elsewhere there is wide consensus "that the geological disposal strategy can be designed and implemented in a manner that is sensitive and responsive to fundamental ethical and environmental considerations." Furthermore, "stepwise implementation of plans for geological disposal leaves open the possibility of adaptation, in the light of scientific progress and social acceptability, over several decades, and does not exclude the possibility that other options could be developed at a later stage."

The problems associated with the final disposal of high-level wastes are neither very large nor complicated. In most respects they are not even novel – the mining industry has been making deep underground excavations since Roman times.

Meanwhile, safe waste management is the well-established norm, disposal technology exists, and full-scale demonstration at acceptable cost can confidently be expected in several countries soon after 2020. Deep geological disposal already exists in the USA for military wastes.

Hazards of wastes
The degree of hazard associated with the separated high-level waste produced from one tonne of used fuel is indicated by Figure 22, and the picture would be similar for used fuel itself except that the plutonium lines would be higher up (around 10 TBq). It can readily be seen that if the minor actinides americium (Am-241, Am-243), neptunium-237 and curium (not shown, but

alpha-decaying to plutonium) were removed during further reprocessing, the radioactivity after a few hundred years would be much reduced.

Two important conclusions can be drawn from the changes in activity with time shown in Figure 22. The first is that the radiological hazard falls by a factor of nearly 1000 between 10 and 1000 years, with relatively little change subsequently. This is because nearly all of the short half-life fission products from the chain reaction will have decayed to negligible levels, leaving behind small quantities of transuranic elements such as americium and neptunium which generally have much longer half-lives. A thousand years is still a long time in human terms, but the object is to put it into stable geological formations where geological time becomes a more meaningful reference (the Earth is 4.6 billion years old). Even the time needed for plutonium to decay away is brief geologically.

The second important point from Figure 22 is that the relative radioactivity of the waste after 1000 years is much the same as that of the corresponding amount of uranium ore. Of course, toxic components of a uranium orebody which outcrops at the surface of the Earth actually do find their way into the biosphere. Waste material in ceramic form buried some 500 metres below the surface in a stable geological structure will have no conceivable chance of doing so. (However this is not to say that surface uranium deposits are dangerous, as the amounts which reach anybody are very small.)

Final disposal

Most countries with nuclear facilities planned or operational have active programs aimed at defining and testing suitable deep geological disposal sites. The aim of this work is to locate areas where multiple barriers can be established between the wastes and the human environment. Some of the barriers, both natural and artificial, are:

- Stable and insoluble form of waste (glass or ceramic).
- Seal in corrosion-resistant containers (e.g. stainless steel, copper).
- In wet rock, pack with bentonite clay to inhibit groundwater movement and insulate from minor earth movement.
- Locate deep underground (e.g. 500 metres deep) in stable rock structure.

A truck carrying TRUPACT-II shipping containers approaching the Waste Isolation Pilot Plant in New Mexico
(US Department of Energy)

Three types of geological structures are being widely studied for this purpose – hard crystalline rocks such as granite, clay beds, and rock salt beds. Suitable locations have been identified in several countries and sites are now undergoing detailed evaluation. Most approaches plan to utilise conventional mining techniques involving shaft-sinking and development of extensive drives and rooms. One purpose-built deep geological repository (Waste Isolation Pilot Plant) has been operating in New Mexico, USA since 1999 but this is only for long-lived military wastes.

The problems involved in carrying out this work are essentially technical. Conventional mining and engineering design techniques together with monitoring of rock temperatures and stresses will enable disposal operations to be carried out to a very high order of safety. The engineering and organisational tasks of maintaining effective isolation of hazardous materials are not new, nor are they peculiar to nuclear wastes.

The question of geological stability of the rock structure is very important for the long-term integrity of the waste repository. There are a number of rock structures which have been stable for more than half of the Earth's 4600 million years, suggesting little likelihood of significant movement for isolation periods of thousands of years.

While deep geological disposal of nuclear wastes is potentially permanent, it is possible to leave open the option of making the material retrievable by future generations. Used fuel will always remain a resource, and it is possible that in a few decades hence, it will have a value which makes it worth recovering for recycling.

The Japanese Cavern Retrievable (CARE) concept involves two distinct stages: storage in ventilated underground caverns with the wastes in overpacks (hence shielded) and fully accessible, followed by backfilling and sealing the caverns after about 300 years. The initial institutional control period – during which most of the radiological decay of the wastes occurs – ensures that the thermal load is much reduced by stage 2, allowing waste casks to be packed more closely together than other with disposal concepts. This is readily adaptable for used fuel, with the overpacks then being the shipping casks.

In Sweden and Finland, sites have been identified for permanent geological repositories, with strong local public support, and construction is under way.

Perspective on wastes

Since even in countries depending substantially on the technology, nuclear wastes generally comprise only about 1% of all toxic industrial wastes, it is relevant to compare their toxicity with that of common industrial poisons used every day by industry and others. Arsenic, of course, is routinely distributed to the environment as a herbicide and in treated timbers. Barium is not uncommon, and chlorine is in widespread domestic and industrial use. Then there are waste mercury compounds, and organochlorines such as polychlorinated biphenyls (PCBs) and hexachlorobenzene which are extremely hazardous – many are also liquids. Considering the quantities available, these are arguably far more hazardous than civil nuclear wastes, which have given rise to no comparable problems or hazards in over 50 years.

Radioactive wastes are treated much more conservatively than other toxic wastes in relation to risks to people and the environment. Most toxic wastes do not break down naturally in a way that corresponds to the progressive decay of radioactivity, and thus they mostly have an infinite life.

There is now no question that disposal of high-level wastes, when it comes of age, will be safe. The wastes, though very toxic when first produced, are small in quantity and no more hazardous in total than other more familiar materials. Nevertheless, they have come to epitomise the 'NIMBY' ('not in my backyard') syndrome of modern society, where the benefits of technology and economic development are readily accepted, but any dirty, unpleasant or fearful aspects, however safe they may actually be, are left to others.

While each country is responsible for disposing of its own wastes of all kinds, the possibility of international nuclear waste repositories is being very actively considered, notably in Europe.

Final disposal costs

The cost of dealing properly with wastes is important. In the USA utility customers pay a fee of 0.1 cent per kilowatt hour (¢/kWh) into a national Nuclear Waste Fund for management and eventual disposal of their used fuel. By mid-2012 this fund had accumulated over $35 billion (less funds spent on Yucca Mountain, so balance of $27.5 billion). This levy is routinely assessed, and found to be adequate. Canadian utilities collect a fee of about 0.1 ¢/kWh to finance future disposal of used fuel. In Sweden a levy of some 0.3 ¢/kWh finances the country's smoothly functioning waste repository for low- and intermediate-level wastes and research on disposal of used fuel. Similar arrangements are in place in other countries, with the expectation that final disposal of all nuclear wastes will be fully funded in advance.

A natural analogue: Oklo

Although highly active wastes from modern nuclear power have not yet been buried for long enough to observe the results, this process has in fact occurred naturally in at least one location. In what is now Gabon in west Africa, about 2 billion years ago, at least 17 natural nuclear reactors commenced operation in a rich deposit of uranium ore at Oklo. Each operated at less than 100 kW thermal. At that time the concentration of U-235 in all natural uranium was some 3.7% instead of 0.7% as at present[7].

These natural chain reactions, started spontaneously and with the presence of water acting as a moderator, continued for about 2 million years before finally dying

away. During this long reaction period about 5.4 tonnes of fission products as well as some 2 tonnes of plutonium together with other transuranic elements were generated at the reactor locations in the orebody. It appears that each reactor operated in pulses of about 30 minutes – interrupted when the water turned to steam thereby switching it off for a few hours until it cooled.

The initial radioactive products have long since decayed into stable elements but close study of the amount and location of these has shown that there was little movement of radioactive wastes during and after the nuclear reactions. Plutonium and the other transuranics remained immobile. This is remarkable in view of the fact that groundwater had ready access to the wastes and they were not in a chemically inert form (such as glass). However, waste materials do not necessarily move freely through the ground even in the presence of water because of their being adsorbed onto clays[8].

Thus the only known 'test' of underground nuclear waste disposal, at Oklo, was successful over a long period in spite of the characteristics of the site. Such a water-logged, sandstone/shale structure would not be considered for disposal of modern toxic wastes, nuclear or otherwise, although the clays and bitumen present played an important part in containing the wastes.

6.5 DECOMMISSIONING NUCLEAR REACTORS

So far 85 commercial nuclear power reactors, 45 experimental or prototype reactors, over 250 research reactors and a number of fuel cycle facilities, have been retired from operation. Of the power reactors, about 15 have been fully dismantled, with the sites released for unconditional use. About 51 of them are being dismantled, 48 are in Safestor (long-term enclosure after partial dismantling), three were entombed, and for others the decommissioning strategy is not yet specified. The broken-up pieces from dismantling are buried along with other intermediate-level wastes[9].

At mid-2012, 11 reactors have been shut down and decommissioned as a result of an accident or serious incident, 25 (in ten countries) have been closed prematurely by political decision, and 97 have closed

---

[7] See also Appendix 2. U-235 decays about six times faster than U-238, whose half-life is about the same as the age of the Earth.
[8] Leaks from the military waste tanks in the USA also demonstrated the ability of clay soils to retain fission products and transuranics.
[9] Many nuclear submarines have also been decommissioned over the last two decades.

having fulfilled their purpose or due to being no longer economic to run. Of the 97, a total of 43 were prototype or experimental reactors; most of the rest ran for their full design life.

The International Atomic Energy Agency (IAEA) has defined three options for decommissioning, after removal of the fuel. These definitions have been internationally adopted:

- Immediate Dismantling (or Early Site Release/Decon in USA). This option allows for the facility to be removed from regulatory control relatively soon after shutdown or termination of regulated activities. Usually, the final dismantling or decontamination activities begin within a few years, depending on the facility. Following removal from regulatory control, the site is then available for re-use.
- Safe Enclosure (or Safstor). This option postpones the final removal of controls for a longer period, usually in the order of 40-60 years. The facility is placed into a safe storage configuration until the radioactivity decays to a low level, after which dismantling and decontamination takes place.
- Entombment. This option entails placing the facility into a condition that will allow the remaining radioactive material to remain on-site without the requirement of ever removing it totally. This option usually involves reducing the size of the area where the radioactive material is located and then encasing the material in a long-lived structure such as concrete, to ensure the remaining radioactivity is finally of no concern.

Each option has its benefits and disadvantages, and national policy with local requirements will determine which approach is adopted. In the case of immediate dismantling with early site release, responsibility for the decommissioning is not transferred to future generations. The experience and skills of operating staff can also be utilised during the decommissioning program. Alternatively, Safe Enclosure or Safstor allows significant reduction in residual radioactivity, thus reducing radiation hazard during the eventual dismantling. The expected improvements in mechanical techniques during this period should also lead to a reduction in both hazards and costs.

In the case of nuclear reactors, about 99% of the radioactivity is associated with the fuel which is removed before adopting any of the three options. Apart from any surface contamination of plant, the remaining radioactivity comes from 'activation products' such as steel components which have long been exposed to neutron irradiation. Their atoms are changed into different isotopes such as iron-55, cobalt-60, nickel-63 and carbon-14. The first two are highly radioactive, emitting gamma rays. However, their half-lives are such that, after 50 years from shutdown, their radioactivity is much diminished and the risk to workers largely gone. Overall, in 100 years after shutdown, the level of radioactivity from activation products falls by a factor of 100,000.

Decommissioning costs

The total cost of decommissioning is dependent on the sequence and timing of the various stages of the program. Deferment of a stage tends to reduce its cost, due to decreasing radioactivity, but this is offset by increased storage and surveillance costs.

Even allowing for uncertainties in cost estimates and applicable discount rates, decommissioning contributes less than 5% to total nuclear electricity generation costs. In the USA, many utilities have revised their cost projections downwards in the light of experience. Financing methods vary from country to country. Among the most common are:

- External sinking fund (nuclear power levy). This is built up over the years from a percentage of the electricity rates charged to consumers. This is the main US system and variants are widely used elsewhere so that sufficient funds are set aside. Proceeds are placed in a trust fund, possibly outside the utility's control, during the reactor's operating lifetime to cover the cost of decommissioning.
- Prepayment, where money is deposited in a separate account to cover decommissioning costs even before the plant begins operation. This may be done in a number of ways but the funds cannot be withdrawn other than for decommissioning purposes.
- Surety fund, letter of credit, or insurance purchased by the utility to guarantee that decommissioning costs will be covered even if the utility defaults.

In the USA, utilities are collecting 0.1 to 0.2 ¢/kWh to fund decommissioning. They must then report regularly to the regulator on the status of their decommissioning trust funds. About two-thirds of the total estimated cost of decommissioning all US nuclear power plants has

[10] OECD/NEA 2003, *Decommissioning Nuclear Power Plants – policies, strategies and costs.*

Decommissioning of the Windscale Advanced Gas-cooled Reactor (WAGR), UK – removal of heat exchangers

already been collected, leaving a liability of about $9 billion to be covered over the remaining operating lives of 104 reactors.

An OECD survey published in 2003[10] reported US dollar costs (at 2001 prices) of decommissioning by reactor type. For western PWRs, most were $200-500/kWe; for VVERs costs were around $330/kWe; for BWRs $300-550/kWe; and for CANDU $270-430/kWe. For gas-cooled reactors the costs were much higher due to the greater amount of radioactive materials involved, reaching $2600/kWe for some UK Magnox reactors. While the costs will have risen since then, there is no great uncertainty involved.

Decommissioning experience

To decommission its retired gas-cooled reactors at the Chinon, Bugey and St Laurent nuclear power stations, Electricité de France chose partial dismantling and Safstor, postponing final dismantling and demolition for 50 years. As other reactors will continue to operate at those sites, monitoring and surveillance do not add to the cost.

Germany chose more rapid direct dismantling over Safe Enclosure for the closed Greifswald nuclear power station in former East Germany, where five reactors had been operating. Similarly, the site of the 100 MWe Niederaichbach nuclear power plant in Bavaria was declared fit for unrestricted agricultural use in mid-1995.

Experience in the USA has varied, but about 12 power reactors are using the Safstor approach, while 15 are using, or have undertaken, immediate dismantling. Procedures are set by the Nuclear Regulatory Commission (NRC).

For Trojan (1180 MWe PWR) in Oregon the dismantling was undertaken by the utility itself. The plant closed in 1993, steam generators were removed, transported and disposed of at Hanford in 1995, and the reactor vessel was removed and transported to Hanford in 1999. Except for the used fuel storage, the site was released for unrestricted use in 2005.

A US immediate dismantling project was the 60 MWe Shippingport reactor, which operated from 1957 to 1982. It was used to demonstrate the safe and cost-effective dismantling of a commercial-scale nuclear power plant and the early release of the site. Defuelling was completed in two years, and five years later in 1989 the site was released for use without any restrictions. Because of its small size, the pressure vessel could be removed and disposed of intact, rather than having to be cut up.

At multi-unit nuclear power stations, the choice has been to place the first closed unit into Safstor until the others end their operating lives, so that all can be decommissioned in sequence. This will optimise the use of staff and the specialised equipment required for cutting and remote operations, and hence achieve cost benefits.

Thus, after 14 years of comprehensive clean-up activities, including the removal of fuel, debris and water from the 1979 accident, Three Mile Island 2 was placed in Safstor until 2014, when the operating licence of unit 1 was originally expected to expire, so that both units might be dismantled together. Safstor was also being used for San Onofre 1, which closed in 1992, until licences for units 2 and 3 expired in 2013, but after regulatory changes, dismantling was brought forward to 1999, so it became an immediate dismantling project.

Immediate dismantling was also the option chosen for Fort St Vrain, a 330 MWe high-temperature gas-cooled reactor which was closed in 1989. This took place on a fixed-price contract for $195 million (hence costing less than 1 ¢/kWh despite a short operating life) and the project proceeded on schedule allowing the nuclear licence to be relinquished early in 1997 – the first large US power reactor to achieve this.

Another such US project was Maine Yankee, a 860 MWe plant which closed down in 1996 after 24 years' operation. The containment structure was finally demolished in 2004 and, except for 5 hectares occupied by the dry store for used fuel, the site was released for unrestricted public use in 2005, on budget and on schedule.

See also WNA information paper *Decommissioning Nuclear Facilities*.

7. Other nuclear energy applications

Other than supplying base-load power to the electricity grid, nuclear fission (and radioactive decay of materials from it) has several other applications. This Chapter describes the role of nuclear energy in some well-established and also emerging areas which could become comparably important to the base-load power role.

There is both immediate and long-term potential for nuclear energy to provide a major contribution to world transport needs through expansion of its present role. Most obviously, there is the current provision of nuclear electricity to run major railway networks in Europe and east Asia, and the future expansion of this as electrification is extended. Here it replaces fossil fuels as a source of the electricity, making those networks a genuinely clean form of transport.

7.1 TRANSPORT

Nuclear power is immediately relevant to road transport and motor vehicles by replacing fossil fuels as a source of the electricity used for charging electric vehicles, which can mean that they are properly zero-emission transport from the source of the electricity through to the actual mobility. Plug-in hybrid and pure electric vehicles use electricity, and especially off-peak base-load power from the grid, for recharging and their increased use will change the pattern of daily demand.

In addition, nuclear heat can be used for production of liquid hydrocarbon fuels from coal (see Section 7.3 below on *Process heat*) and also for biofuels. And further out, hydrogen for oil refining and for future fuel cell vehicles may be made electrolytically and, eventually, thermochemically using high-temperature reactors (see Section 7.2 below on *Hydrogen*).

Full electric vehicles (EV) already have a niche market in several countries, but have been limited by battery capacity and weight. Improved battery technology greatly increases their potential. Hybrid vehicles powered by batteries which are largely charged by an internal combustion (IC) engine are now very popular, and the development from these – plug-in hybrid electric vehicles (PHEV) – will enable most of the charging to be from the electricity grid.

Widespread use of PHEVs and EVs – which get much of their energy from the electricity grid overnight at off-peak rates – will increase electricity demand by about 15%, but more significantly it will mean that a greater proportion of the electricity can be generated by base-load plant, and hence at lower average cost. There is likely to be a 30-40% increased requirement for base-load capacity, and a corresponding decrease in intermediate- and peak-load plant. This has the potential to lower all electricity costs significantly (base-load power being lower-cost per kWh), and where those base-load plants are nuclear, the power will also be emissions-free (see Figure 5A in Chapter 2).

Partnerships are starting to emerge between power and automotive companies in anticipation of wider use of PHEVs and EVs in Europe. Deploying the means of charging them is more of a challenge where most cars are not garaged overnight so must be charged elsewhere, often more rapidly. Apart from the technical challenges in the car, recharging systems and protocols for billing need to be developed.

Fuel cell bus with six roof mounted hydrogen fuel tanks

Battery technology is the key for both PHEV and EV. Batteries should achieve high capacity, have low mass and low cost, coupled with safety and a long life, and be capable of repeated deep discharge.

While current automotive fuels provide 12-14 MJ per kilogram mass (net of IC engine efficiency, or 45 MJ/kg gross), the best batteries provide only 2-3 MJ/kg (550-800 Wh/kg net), and that at twice the volume. Commercial batteries are much less than this (see below). Lead-acid batteries are well known for starting cars and running accessories, as well as for the propulsion of older electric vehicles. But they are very heavy and only last a few years. Nickel metal hydride (NiMH) batteries are well-proven and reasonably durable, though can be damaged under some discharge conditions. They are similar to nickel cadmium (NiCd) batteries, but use a hydrogen-absorbing alloy as the cathode instead of cadmium.

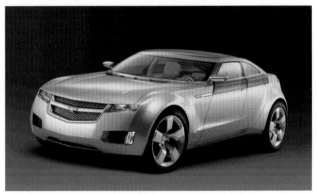

GM Volt plug-in hybrid car

Lithium-ion batteries deliver more power from less mass and are constantly being improved in relation to safety, reliability and durability. Newer ones use manganese oxides or iron phosphate cathodes, which are reliable. Lithium-ion batteries hold about 100-200 watt hours per kilogram of battery mass, the much safer and more durable lithium-ion iron phosphate batteries being at the lower end of this range. Costs are still around $1000 per kWh capacity.

Nissan, EDF, and others envisage an infrastructure integrating three types of charging systems: from household supply overnight (6-8 hours, off-peak), similar slower charge in parking lots during the day, and fast charging points which will give an 80% charge in 30 minutes. In addition to these there should be 5-minute battery pack changeovers for long trips, raising the possibility of batteries being leased rather than owned, or electricity suppliers selling a range of services configured for different users, not just batteries and power. Ford is exploring intelligent vehicle-grid systems in the USA.

Focusing on the home base, using a 13 amp plug and 240 volt system (as standard in the UK), a 16 kWh battery pack such as in the GM Volt could be recharged in 5.5 hours. Many battery packs will be much larger than this, so 40 amp charge points may often be necessary for overnight charging, particularly with 110 volt systems.

See also WNA information paper *Electricity and Cars*.

7.2 HYDROGEN PRODUCTION AND USE

Hydrogen is already a significant chemical product, about half of annual production being used in making nitrogen fertilisers via the Haber process and about half to convert low-grade crude oils (including those from tar sands) into transport fuels[1]. Both uses are increasing significantly. Some is used for other chemical processes. World consumption in 2011 was about 80 million tonnes[2], growing at about 7% per year. There is a lot of experience handling it on a large scale.

About 96% of hydrogen is made from fossil fuels: half from natural gas, 30% from liquid hydrocarbons and 18% from coal. This gives rise to quantities of carbon dioxide emissions – each tonne produced gives rise to 11 tonnes of CO_2. Electrolysis accounts for only 4%.

Like electricity, hydrogen is an energy carrier (but not a primary energy source). As oil becomes more expensive, hydrogen may replace it as a transport fuel and in other applications. This development becomes more likely as fuel cells are developed, with hydrogen as the preferred fuel. If natural gas also becomes expensive, or constraints are put on carbon dioxide emissions, non-fossil sources of hydrogen will become necessary.

As with electricity, hydrogen for transport use will tend to be produced near where it is to be used. This will have

---

[1] *e.g.* $(CH)_n$ from tar sands or $(CH_{1.5})_n$ heavy crude to $(CH_2)_n$ transport fuel.
[2] In thermal terms @ 121 MJ/kg: 9680 PJ, about the same as the thermal equivalent of US nuclear electricity.

major geo-political implications as industrialised countries become less dependent on oil and gas from distant parts of the world.

In the short term, hydrogen can be produced economically by electrolysis of water in off-peak periods, enabling much greater utilisation of base-load plants, and in particular nuclear plants. In future, the heat from high-temperature reactors is likely to be used directly for thermochemical production of hydrogen.

In the USA, producing 11 million tonnes of hydrogen per year (with a thermal energy content of 48 GWt) consumes 5% of US natural gas usage. The use of hydrogen for all US transport would require some 200 million t/yr of hydrogen.

All this points to the fact that while a growing hydrogen economy already exists, linked to the worldwide chemical and refining industry, a much greater one is in sight. With new uses for hydrogen as a fuel, the primary energy demand for its production may approach that for electricity production.

Nuclear power already produces electricity as a major energy carrier. It is well placed to produce hydrogen if this becomes a major energy carrier also.

The evolution of nuclear energy's role in hydrogen production over perhaps three decades is seen to be:

- Electrolysis of water, using off-peak capacity.
- Use of nuclear heat to assist steam reforming of natural gas, which is energy intensive and requires temperatures of up to 900°C. This well-established process however has carbon dioxide as a waste product.
- High-temperature electrolysis of steam at over 800°C, using heat and electricity from nuclear reactors. There has been demonstration of this at laboratory scale in USA.
- High-temperature thermochemical production using nuclear heat. Several direct thermochemical processes are being developed for producing hydrogen from water.

The economics of thermochemical hydrogen production depend on the efficiency of the method used. For economic production, high temperatures – over 900°C – are required to ensure rapid throughput and high conversion efficiencies. Efficiency of the whole process (original heat to thermal content of hydrogen) then moves from about 25% with today's reactors driving electrolysis, to 36% with more efficient reactors doing so, to 45%

Installing a hydrogen fuel tank in the Honda FCX Clarity

for high-temperature electrolysis of steam, to about 50% or more with direct thermochemical production. Combined cycle plants producing both hydrogen and electricity may reach efficiencies of 60%.

Hydrogen from nuclear heat

Each of the leading thermochemical processes relies on the decomposition of sulfuric acid to produce oxygen and sulfur dioxide. This is endothermic (heat absorbing) and high temperatures (800-1000°C) are needed for it to proceed efficiently:

$$H_2SO_4 \rightarrow H_2O + SO_2 + \tfrac{1}{2}O_2 \text{ (at low pressure)}$$

There are then several possibilities.

In the iodine-sulfur (IS) process, iodine combines with the SO_2 and water to produce hydrogen iodide and sulfuric acid. This is known as the Bunsen reaction and is exothermic (heat releasing), occurring at low temperature (120°C):

$$I_2 + SO_2 + 2H_2O \rightarrow 2HI + H_2SO_4$$

The HI then dissociates to hydrogen and iodine at about 350°C, endothermically:

$$2HI \rightarrow H_2 + I_2$$

This can deliver hydrogen at high pressure.

Combining all this, the net reaction is then:

$$H_2O \rightarrow H_2 + \tfrac{1}{2}O_2$$

All the reagents other than water are recycled – there are no effluents. The inputs are heat and water, the outputs are hydrogen and oxygen, and the oxygen by-product also has value.

The Japan Atomic Energy Authority (JAEA) has demonstrated laboratory-scale hydrogen production with this process, at a rate of up to 35 litres/hr. The Sandia National Laboratory in the USA and the French CEA are also developing the process with a view to using high-temperature reactors for it.

Production reactors and plans

High temperature (750-1000°C) is required for thermochemical hydrogen production, though at 1000°C the conversion efficiency is three times that at 750°C. The chemical plant needs to be isolated from the nearby reactor, for safety reasons, possibly using an intermediate helium or molten fluoride loop[3].

Two potentially suitable reactor concepts have been identified:

- High-temperature gas-cooled reactor (HTR), either the pebble bed or hexagonal fuel block type, with helium coolant at high pressure. Modules of up to 600 MWt will operate at 950°C but can be hotter (see also Section 4.5).
- Advanced high-temperature reactor (AHTR), a modular reactor using a coated-particle graphite-matrix fuel and with molten fluoride salt as primary coolant. This is similar to the HTR but operates at low pressure (<1 atmosphere) and higher temperature, and gives better heat transfer. Sizes of 1000 MWe/2000 MWt are envisaged.

A 600 MWt plant would produce about 200 tonnes of hydrogen per day, which is well matched to the scale of current industrial demand for hydrogen.

The Korean Atomic Energy Research Institute (KAERI) has submitted a very high temperature reactor (VHTR) gas-cooled design to the Generation IV International Forum[4] with a view to hydrogen production from it. This is envisaged as 300 MWt modules each producing 30,000 tonnes of hydrogen per year. KAERI envisages operation in the 2020s.

The Japan Atomic Energy Agency (JAEA) expects to confirm the safety of high-temperature reactors and establish operational technology for an iodine-sulfur (IS) plant to make hydrogen thermochemically. In 2004, a coolant outlet temperature of 950°C was achieved in its High-Temperature Engineering Test Reactor (HTTR) – a world first, and opening the way for thermochemical hydrogen production. By 2015, an IS plant producing 1000 m^3/hr (90 kg/hr, 2t/day) of hydrogen should be linked to the HTTR to confirm the performance of an integrated production system.

In the meantime, hydrogen can be produced by electrolysis of water, using electricity from any source. Non-fossil sources, including intermittent ones such as wind and solar, are important possibilities (thereby solving a problem of not being able to store the electricity from those sources). However, the greater efficiency of electrolysis at high temperatures favours a nuclear source for both heat and electricity.

Hydrogen as a fuel

Burning hydrogen produces only water vapour, with no carbon emissions.

In motor vehicles, hydrogen can be burned in a normal internal combustion engine, and some test cars are thus equipped. Trials in aircraft have also been carried out. However, its main use is likely to be in fuel cells. A fuel cell is conceptually a refuelable battery, making electricity as a direct product of a chemical reaction. But where normal batteries have all the active ingredients built in at the factory, fuel cells are supplied with fuel from an external source.

Fuel cells catalyse the oxidation of hydrogen directly to water at relatively low temperatures while producing electricity. The claimed theoretical efficiency of converting chemical to electrical energy is about 60% (or more). However, in practice about half that has been achieved, except for the higher-temperature solid oxide fuel cells, which can achieve 46% efficiency.

On-board storage is the principal problem for hydrogen as an automotive fuel – it is impossible to store it as simply and compactly as gasoline or LNG[5] fuel. The options are

---

[3] Molten fluoride salts are a preferred interface fluid between the nuclear heat source and the chemical plant. The aluminium smelting industry provides substantial experience in managing them safely.

[4] The Generation IV International Forum (GIF) is an international collective representing governments of 13 countries collaboratively developing seven (originally six) reactor designs for deployment in the 2030s.

[5] Liquefied natural gas.

to store it at very low temperature (cryogenically), at high pressure, or chemically as hydrides. The last is seen to have most potential, though refuelling a vehicle is less straightforward. Pressurised storage is the main technology available now and this means that at 345 times atmospheric pressure (34.5 MPa, 5000 psi), ten times the volume is required than for an equivalent amount of petrol/gasoline. This disadvantage is coupled with a weight penalty due to the storage system, which is about 50 times heavier than the hydrogen it stores – the aim is to get it down to 20 times as heavy in the next few years, and perhaps ten times as heavy one day.

The first hydrogen fuel cell electric cars are expected to be on the fleet market in the next decade. Japan has a goal of having 5 million fuel cell vehicles on the road by 2020, though it may be that electric vehicles using batteries are sufficiently successful to displace this.

At present, fuel cells are much more expensive to make than internal combustion engines (burning petrol/gasoline, LPG[6], natural gas or hydrogen). Current fuel cell design consists of bipolar plates in a frame, and the developer of the proton exchange membrane type, Dr Geoffrey Ballard, suggested that a new geometry is required to bring the cost down and make the technology more widely available to a mass market. Other assessments point out that fuel cells are intrinsically not simple and there are no obvious reasons to expect them to become cheap.

The initial use of hydrogen for transport is likely to be municipal bus and truck fleets, and prototypes are already on the road in many parts of the world. These are centrally fuelled, so avoid the need for a retail network, and on-board storage of hydrogen is less of a problem than in cars.

Hydrogen can also be used for stand-alone small-scale stationary generating plants using fuel cells – where higher temperature operation (e.g. of solid oxide fuel cells) and hydrogen storage may be less of a problem or where it is reticulated like natural gas.

An electricity system based on nuclear power could produce hydrogen at a steady rate and store it underground so that it was used in large banks of fuel cells (e.g. 1000 MWe) at peak demand periods each day. Efficiency would be enhanced if by-product oxygen instead of air were used in the fuel cells.

Hydrogen for agricultural fertilisers

According to Norman Borlaug, 1970 Nobel laureate and 'grandfather of the green revolution', organic nitrogen compounds present in the world's soils are only sufficient to feed one-third of today's population. The rest must come from inorganic additions. Most of the world's nitrogen fertilisers are made using the Haber[7] process, combining abundant atmospheric nitrogen with hydrogen. The resulting ammonia is then oxidised to nitrates. But the hydrogen has to be made from fossil fuels, principally methane, i.e. natural gas. This is costly and it gives rise to substantial carbon dioxide emissions[8].

If nuclear energy is used to make the hydrogen from water, the carbon dioxide is avoided and a valuable organic chemistry feedstock is conserved. An abundant supply of low-cost hydrogen would greatly boost world agricultural productivity through increased availability of nitrogen fertilisers.

Hydrogen for oil refining

A major use of hydrogen today is hydrogenation of heavy crude oil, a process that breaks down the long-chain hydrocarbons to yield synthetic crude oil (about 5 kg of hydrogen is used per barrel). Steam reforming of natural gas is used to produce the hydrogen feedstock.

Nuclear power could make this steam and electricity, and use some of the electricity in high-temperature electrolysis for hydrogen production. (Heavy water and oxygen could be valuable by-products of electrolysis.)

See also WNA information paper *Nuclear Process Heat for Industry*.

---

[6] Liquefied petroleum gas.

[7] German scientist Fritz Haber invented the process in 1909 and received the Nobel Prize for chemistry in 1918 for creating "an exceedingly important means of improving the standards of agriculture and the well-being of mankind", which now looks like a considerable understatement.
$$N_2 + 3H_2 \rightarrow 2NH_3$$
The Haber process produces about 100 million tonnes of nitrogen fertiliser per year and consumes about 3-5% of the world's natural gas production. It accounts for about half of the hydrogen produced annually.

[8] In several steps, but overall: $CH_4 + O_2 \rightarrow CO_2 + 2H_2$

7.3 OTHER PROCESS HEAT

Nuclear process heat has great potential for other end uses than those covered in Sections 7.1 and 7.2, using today's technology.

Recovery of oil from tar sands

From about 2003, various proposals have been made to use nuclear power to produce steam for extraction of oil from Alberta's northern oil sand (tar sand) deposits and also electricity for the major infrastructure involved. At present, a lot of Canada's natural gas is used as an energy source to make steam to liquefy the bitumen, enabling its separation, and to generate electricity for mining and treatment. This amounts to some 30 cubic metres of gas (1100 MJ) per barrel of oil.

With projections of three million barrels per day by 2016, a great deal of gas will be required and the cost exposure is increasing dramatically. The Canadian Energy Research Institute predicts a tripling in gross bitumen production from 2008 to 2020. In fact, Canadian natural gas is inadequate to supply the anticipated expansion in oil sands

Sasol's Secunda Operations in South Africa provide 30% of the country's oil, from coal (Sasol)

output and its use has major carbon dioxide (CO_2) implications which are creating public concern – about 20% of the energy in the oil is required to produce it and about 80 kg of CO_2 per barrel is released.

One proposal from Energy Alberta suggested that a single CANDU 6 reactor (about 1800 MWt) configured to produce 75% steam and 25% electricity would replace 6 million cubic metres per day of natural gas and support production of 175-200,000 barrels per day of oil, saving the emission of much CO_2.

The main difference between natural gas and nuclear steam generation is that a fuel-intensive process is replaced by a capital-intensive one.

Coal to liquids

Coal can be used to make synthetic oil. The Fischer-Tropsch process was originally developed in Germany in the 1920s, and provided much of the fuel for Germany during the Second World War. It then became the basis for oil production in South Africa by Sasol, which now supplies about 30% of that country's gasoline and diesel fuel. The process is a significant user of hydrogen, using carbon monoxide in a catalytic reaction.

The hydrogen is now produced with the carbon monoxide by coal gasification. The carbon monoxide then undergoes the water gas shift reaction to produce more hydrogen[9]. A nuclear source of hydrogen coupled with nuclear process heat would triple the amount of liquid hydrocarbons from the coal and eliminate most CO_2 emissions from the process. Today, using simply black coal, 14,600 tonnes of coal is gasified to produce 25,000 barrels of synfuel 'oil' (with over 25,000 tonnes of CO_2). A proposed hybrid system uses nuclear electricity to electrolyse water for the hydrogen. Only about 4400 tonnes of coal is then required, using oxygen from the electrolysis to produce carbon monoxide which is fed to the Fischer-Tropsch plant with the hydrogen to produce the same 25,000 barrels of synfuel 'oil'. Very little CO_2 results, and this is recycled to the gasifier anyway.

There is considerable potential to extend world oil supplies, with negligible carbon footprint in their production, by this means.

See also WNA information paper on *Nuclear Process Heat for Industry.*

---

[9] $CO + H_2O$ gives CO_2 & H_2

7.4 DESALINATION

It is estimated that one-fifth of the world's population does not have access to safe drinking water, and that this proportion will increase due to population growth relative to water resources. The worst-affected areas are the arid and semi-arid regions of Asia and North Africa. Wars over access to water, not simply energy and mineral resources, are conceivable.

Fresh water is a major priority in sustainable development. Where it cannot be obtained from streams and aquifers, desalination of seawater or of mineralized groundwater is required.

Reverse osmosis desalination, Jersey, UK. The whole plant produces 6000 m^3 of fresh water per day from sea water, using 2 MWe. (Jersey Water)

Most desalination today uses fossil fuels, and thus contributes to increased levels of greenhouse gases. Total world capacity is about 40 million m^3/day of potable water, in some 15,000 plants. Most of these are in the Middle East and North Africa, using distillation processes. The largest plant produces 820,000 m^3/day. Two-thirds of the capacity is processing sea water, and one-third uses brackish artesian water.

The major technology in use and being built today is reverse osmosis (RO) driven by electric pumps which force water through a membrane against its osmotic pressure[10]. A thermal process, multi-stage flash (MSF) distillation using steam, was earlier prominent and it is capable of using waste heat from power plants. With brackish water, RO is much more cost-effective, though MSF gives purer water than RO. A minority of plants use multi-effect distillation (MED) or multi-effect vapour compression (MVC). MSF-RO hybrid plants exploit the best features of each technology for different quality products.

Desalination is energy-intensive. Reverse osmosis needs up to 6 kWh of electricity per cubic metre of seawater (depending on its salt content), hence 1 MWe will produce about 160 m^3 per hour. MSF and MED require heat at 70-130°C and use more electricity, though a newer version of MED (MED-MVC) is reported to be competitive with RO.

A variety of low-temperature heat sources may be used, including solar energy. For brackish water and reclamation of municipal wastewater, RO requires only about 1 kWh/m^3. The choice of process generally depends on the relative economic values of fresh water and particular fuels, and whether cogeneration is a possibility.

Small and medium-sized nuclear reactors are suitable for desalination, often with cogeneration of electricity using low-pressure steam from the turbine and hot seawater feed from the final cooling system. The main opportunities for nuclear plants have been identified as the 80-100,000 m^3/day and 200-500,000 m^3/day ranges.

Some 10% of Israel's water is desalinated, and one large RO plant provides water at 50 cents (US) per cubic metre. Malta gets two-thirds of its potable water from RO. Singapore in 2005 commissioned a large RO plant supplying 136,000 m^3/day – 10% of needs, at 49 cents (US) per cubic metre and it is building a new one double the size to produce water at three-quarters that cost. China's State Council has announced that it aims to have 2.2 to 2.6 million m^3/day seawater desalination capacity operating by 2015.

Nuclear experience

Much relevant experience comes from nuclear plants in Russia, Eastern Europe and Canada where district heating is a by-product. The feasibility of integrated nuclear desalination plants has been proven with over 150 reactor years of experience, chiefly in Kazakhstan, India and Japan. Large-scale deployment of nuclear desalination on a commercial basis will depend primarily on economic factors. Indicative costs are 70-90 cents (US) per cubic metre, much the same as fossil-fuelled plants in the same areas.

---

[10] About 27 bar — therefore RO needs compression of much more than this.

One obvious strategy is to use all the electricity from power reactors (which tend to run at full capacity) to meet grid load when that is high and part of it to drive pumps for RO desalination when the grid demand is low.

The BN-350 fast reactor at Aktau, in Kazakhstan, successfully supplied up to 135 MWe of power while producing 80,000 m^3/day of potable water over some 27 years, about 60% of its power being used for heat and desalination. Although the plant was designed as 1000 MWt, it never operated at more than 750 MWt; however, it established the feasibility and reliability of such cogeneration plants. (In fact, oil/gas boilers were used in conjunction with it, and total desalination capacity through ten MED units was 120,000 m^3/day.)

In Japan, some ten desalination facilities linked to pressurised water reactors operating for electricity production have yielded 1000-3000 m^3/day each of potable water, and over 100 reactor years of experience have accrued. MSF was initially employed, but MED and RO have been found more efficient there. The water is used for the reactors' own cooling systems.

India has been engaged in desalination research since the 1970s and in 2002 set up a demonstration plant coupled to twin 170 MWe PHWR units at the Madras Atomic Power Station, Kalpakkam, in southeast India. This nuclear desalination demonstration project is a hybrid RO/MSF plant, the RO with 1800 m^3/day capacity and the MSF 4500 m^3/day costing about 25% more. They incur a 4 MWe loss in power from the plant. In 2009 a 10,200 m^3/day MVC plant was set up at Kudankulam to supply fresh water for the new plant.

China Guangdong Nuclear Power has commissioned a 10,080 m^3/day desalination plant at its new Hongyanhe nuclear plant at Dalian in the northeast.

New projects

Russia has designed a nuclear desalination project using dual barge-mounted KLT-40 marine reactors (each 150 MWt) and Canadian RO technology to produce potable water, but this has yet to take shape.

South Korea has designed a small nuclear reactor design for cogeneration of electricity and potable water. The 330 MWt SMART reactor (an integral PWR) has a long design life and needs refuelling only every three years. Another concept has the SMART reactor coupled to four MED units, each with a thermal vapour compressor and producing a total of 40,000 m^3/day. Argentina has designed the integral

100 MWt CAREM reactor suitable for cogeneration or desalination alone and is building a prototype.

Spain, the UK, China, India, Pakistan, Egypt, Algeria, Morocco and Tunisia all have projects to build new desalination plants, and the feasibility studies for these often involve nuclear power. Most or all these have requested technical assistance from the UN's International Atomic Energy Agency (IAEA) under its technical cooperation project on nuclear power and desalination. A coordinated IAEA research project initiated in 1998 and involving more than 20 countries reviewed reactor designs intended for coupling with desalination systems as well as advanced desalination technologies.

See also WNA information paper on *Desalination*.

7.5 MARINE PROPULSION

About 180 nuclear reactors power nearly 150 ships, mostly naval vessels and more than 12,000 reactor years of marine operation has been accumulated. Most of the ships are submarines, but they range from ice-breakers to aircraft carriers. Nuclear power is particularly suitable for vessels which need to be at sea for long periods without refuelling, as well as for powerful submarine propulsion.

Work on nuclear powered ships and submarines started in the 1940s, and the first test reactor started up in the USA in 1953. The first nuclear-powered submarine, *USS Nautilus*, put to sea in 1955, marking the transition of submarines from slow underwater vessels to warships capable of sustaining 20-25 knots, submerged for weeks on end. *Nautilus* led to the parallel development of further (*Skate class*) submarines, powered by single pressurised water reactors (PWRs), and an aircraft carrier, *USS Enterprise*, powered by eight PWR units in 1960. A cruiser, *USS Long Beach*, followed in 1961 and was powered by two of these early units. Remarkably, the *Enterprise* remains in service (with new reactors).

By 1962, the US Navy had 26 operational nuclear submarines and 30 under construction. Nuclear power had revolutionised the Navy. The technology was shared with Britain, while French, Russian and Chinese developments proceeded separately.

After the *Skate* class vessels, reactor development proceeded, and in the USA a single series of standardised designs came from both Westinghouse and General Electric, one reactor powering each vessel. Rolls-Royce

Nuclear submarines are able to maintain high speeds underwater for long periods

Compared with the excellent safety record of the US nuclear navy, early Soviet endeavours resulted in a number of serious accidents – five where the reactor was irreparably damaged, and more resulting in radiation leaks. There were more than 20 radiation fatalities. However, by the third generation of marine PWRs in the late 1970s, safety and reliability had become a high priority.

At the end of the Cold War, in 1989, there were over 400 nuclear-powered submarines operational or being built. Some 300 of these submarines have now been scrapped and some on order cancelled, due to weapons reduction programs. Russia and the USA had over 100 each in service, with the UK and France fewer than 20 each and China six. India launched its first in 2009, based on the Russian *Akula-I* class.

France has a nuclear-powered aircraft carrier and ten nuclear submarines. The UK has 12 submarines, all nuclear powered. China is understood to have about ten nuclear submarines. The Russian Navy appears to have about 21 nuclear submarines and has logged over 6000 nautical reactor years.

Civil vessels

Nuclear propulsion has proven technically and economically essential in the Russian Arctic where operating conditions are beyond the capability of conventional ice-breakers. The power levels required for breaking ice up to 3 metres thick, coupled with refuelling difficulties for other types of vessels, are significant factors. The nuclear fleet has increased Arctic navigation from 2 to 10 months per year, and in the Western Arctic, it is year-round.

The ice-breaker Lenin was the world's first nuclear-powered surface vessel (20,000 dwt) commissioned in 1959. It remained in service for 30 years, being retired due to the hull being worn thin from ice abrasion, though new reactors were fitted in 1970. It led to a series of larger ice-breakers, the six 23,500 dwt *Arktika* class, launched from 1975. These powerful vessels have two 171 MWt reactors delivering 54 MW of power at the propellers and are used in deep Arctic waters. The *Arktika* was the first surface vessel to reach the North Pole, in 1977. The seventh and largest *Arktika* class ice-breaker – *50 Years of Victory (50 Let Pobedy)* – entered service in 2007 (12 years later than the 50-year anniversary of 1945 it was to commemorate). It is 25,800 dwt.

For use in shallow waters such as estuaries and rivers, two shallow-draught *Taymyr* class ice-breakers of 18,260 dwt with one reactor delivering 35 MW were built in

built similar units for British Royal Navy submarines and then developed the design further, to its PWR2 reactor.

The USA has the main navy with nuclear-powered aircraft carriers, while both it and Russia have had nuclear-powered cruisers (USA 9, Russia 4). The US Navy has accumulated about 6200 reactor years of experience without any reactor accidents, and operated 82 nuclear-powered ships (11 aircraft carriers, 71 submarines) with 103 reactors as of March 2010. All US aircraft carriers and submarines are nuclear-powered.

Russia built 248 nuclear-powered submarines and five naval surface vessels powered by 468 reactors between 1950 and 2003. Around 50 of these vessels are still in operation. It developed both PWR and lead-bismuth cooled reactor designs, the latter not persisting. Eventually four generations of submarine PWRs were utilised, the last entering service in 1995 in the *Severodvinsk* class.

The largest submarines are the 26,500 tonne Russian *Typhoon* class, powered by twin 190 MWt PWR reactors, though these were superseded by the 24,000 tonne *Oscar-II* class (e.g. *Kursk*) with the same power plant.

Finland, and then fitted with their nuclear steam supply system in Russia and launched from 1989.

Development of nuclear merchant ships began in the 1950s but on the whole has not been commercially successful. The 22,000 tonne US-built *NS Savannah* was commissioned in 1962 and decommissioned eight years later. It was a technical success, but not economically viable. It had a 74 MWt reactor delivering 16.4 MW to the propeller. The German-built 15,000 tonne *Otto Hahn* cargo ship and research facility sailed some 650,000 nautical miles on 126 voyages in 10 years without any technical problems. It had a 36 MWt reactor delivering 8 MW to the propeller. However, it proved too expensive to operate and in 1982 it was converted to diesel.

The 8000 tonne Japanese *Mutsu* was the third civil vessel, put into service in 1970. It had a 36 MWt reactor delivering 8 MW to the propeller. It was dogged by technical and political problems and was an embarrassing failure. These three vessels used reactors with low-enriched uranium fuel (3.7-4.4% U-235).

In 1988, the *NS Sevmorput* was commissioned in Russia, mainly to serve northern Siberian ports. It is a 61,900 tonne LASH carrier (taking lighters to ports with shallow water) and container ship with ice-breaking bow. It is powered by the same KLT-40 reactor as used in larger ice-breakers, delivering 32.5 propeller MW from the 135 MWt reactor, and it needs refuelling only every 15 years.

Russian experience with nuclear-powered Arctic ships total about 300 reactor years to the end of 2008. A more powerful ice-breaker of 110 MW at the propellers and 55,600 dwt is planned, with further dual-draught ones of 32,400 dwt and 60 MW power.

With increasing attention being given to greenhouse gas emissions arising from burning fossil fuels for international air and marine transport, and the excellent safety record of nuclear-powered ships, it is likely that there will be renewed interest in marine nuclear propulsion for civil use. Economic constraints on fossil fuel use in transport will help this reconsideration. The world's merchant shipping is reported to have a total power capacity of 410 GWt, about one-third that of world nuclear power plants.

In 2010 Babcock International's marine division completed a study on developing a nuclear-powered LNG tanker (which requires considerable auxiliary power as well as propulsion). The study indicated that particular routes and

Large Russian nuclear icebreaker Yamal – 23,500 tonne dwt

cargoes lent themselves well to the nuclear propulsion option, and that technological advances in reactor design and manufacture had made the option more appealing.

Also in 2010, Lloyd's Register embarked upon a two-year study with US-based Hyperion Power Generation (now Gen4 Energy), British vessel designer BMT Group, and Greek ship operator Enterprises Shipping and Trading SA "to investigate the practical maritime applications for small modular reactors. The research is intended to produce a concept tanker-ship design," based on a 70 MWt reactor such as Gen4 Energy's. The project includes research on a comprehensive regulatory framework led by the International Maritime Organisation (IMO), and supported by the International Atomic Energy Agency (IAEA) and regulators in the countries involved.

Nuclear propulsion systems

Naval reactors (with one exception) have been pressurised water types, which differ from commercial reactors producing electricity in that:

- They need to deliver a lot of power from a very small volume and therefore run on highly-enriched uranium (over 90% in US submarines, c. 20-25% in some Western vessels, and up to 45% in later Russian ones).
- The fuel is not UO_2 but a uranium-zirconium or uranium-aluminium alloy (c. 15% U with 93% enrichment, or more U with less enrichment) or a metal-ceramic composite.
- They have long core lives, so that refuelling is needed only after 10 or more years, and new cores are designed to last 50 years in carriers and 30-40 years in submarines (e.g. *US Virginia class*).
- The design enables a compact pressure vessel while maintaining safety. The *Sevmorput* pressure vessel for a relatively large marine reactor is 4.6 m high and 1.8 m diameter, enclosing a core 1 m high and 1.2 m diameter.

The long core life is enabled by the relatively high enrichment of the uranium and by incorporating a 'burnable poison' such as gadolinium – which is progressively depleted as fission products and actinides accumulate. These would normally cause reduced fuel efficiency, but the two effects cancel one another out. Long-term integrity of the compact reactor pressure vessel is maintained by providing an internal neutron shield.

Reactor power ranges from 10 MWt (in a prototype) up to 200 MWt in the larger submarines and 300 MWt in surface ships such as the *Kirov* class battle cruisers. The French Rubis class submarines have a 48 MWt reactor which needs no refuelling for 30 years. Russia's *Oscar-II* class has two 190 MWt reactors.

The Russian, US and British navies rely on steam turbine propulsion, the French and Chinese use the turbine to generate electricity for propulsion. Russian ballistic missile submarines as well as all surface ships since the *USS Enterprise* are powered by two reactors. Other submarines are powered by one.

Cassini's RTGs undergo mechanical and electrical verification testing

The larger Russian ice-breakers use two KLT-40 nuclear reactors with 45-75% enriched fuel and 3-4 year refuelling interval. They drive steam turbines and each delivers up to 33 MW at the propellers. The large freighter *Sevmorput* uses one of the same units, though with 90% enriched fuel.

For the next generation of Russian ice-breakers, integrated light water reactor designs are being investigated possibly to replace the conventional PWR.

Dismantling decommissioned nuclear-powered submarines has become a major task for US and Russian navies. After defuelling, normal practice is to cut the reactor section from the vessel for disposal in shallow land burial as low-level waste. In Russia the whole vessels, or the sealed reactor sections, sometimes remain stored afloat indefinitely, though Western-funded programs are addressing this and all decommissioned submarines were due to be dismantled by 2012.

See also WNA information paper *Nuclear Powered Ships*.

7.6 RADIOISOTOPE SYSTEMS AND REACTORS FOR SPACE

There have been periodical bursts of interest in the use of nuclear fission power for space missions, but these have lapsed due to low priority in research budgets.

While Russia has used over 30 fission reactors in space, the USA has flown only one – the SNAP-10A (System for Nuclear Auxiliary Power) in 1965.

From 1959 to 1973 a US nuclear rocket program – Nuclear Engine for Rocket Vehicle Applications (NERVA) – focused on nuclear power replacing chemical rockets for the latter stages of launches. NERVA used graphite-core reactors to heat hydrogen, which was then expelled through a nozzle. Some 20 engines were tested in Nevada and yielded thrust up to more than half that of the space shuttle launchers. Since then, 'nuclear rockets' have been about space propulsion, not launches.

Radioisotope systems

Radioisotope power sources have been an important source of energy in space since 1961. Radioisotope thermoelectric generators (RTGs) have been the main power source for US space work for nearly 50 years, since 1961. The high decay heat of plutonium-238 (0.56 W/g) enables its use as an electricity source in the RTGs of spacecraft, satellites, navigation beacons *etc.*, and its alpha decay process calls for minimal shielding. Heat from the oxide fuel is converted to electricity through static thermoelectric elements (solid-state thermocouples), with no moving parts. RTGs are safe, reliable, and maintenance-free and can provide heat or electricity for decades under very harsh conditions, particularly where solar power is not feasible.

So far 45 RTGs have powered 25 US space vehicles including *Apollo, Pioneer, Viking, Voyager, Galileo* and *Ulysses* space missions as well as many civil and military satellites. The *Cassini* spacecraft carries three RTGs providing 870 watts of power as it explores Saturn. The *Voyager* spacecraft, which have sent back pictures of distant planets, have already operated for 30 years and are expected to send back signals powered by their RTGs from 13 billion kilometres for another ten years or so. The *Viking* and *Rover* landers on Mars in 1975 depended on RTG power sources, as does the 900 kg Mars Science Laboratory Rover launched in 2011 and due to arrive in August 2012. (The two Mars Exploration Rovers operating 2004-09 used solar panels and batteries.)

The latest RTG is a 290-watt system known as the GPHS-RTG (General Purpose Heat Source Radioisotope Thermoelectric Generator), the thermal power for this system being provided by 18 GPHS units. Each GPHS contains four iridium-clad Pu-238 fuel pellets, stands 5 cm tall, 10 cm square and weighs 1.44 kg. The Multi-Mission RTG (MMRTG) will use 8 GPHS units producing 2 kW thermal, which is used to generate 110 watts of power for the Mars Science Laboratory, a large mobile laboratory – the rover *Curiosity*, which at 890 kg is about five times the mass of previous Mars rovers.

The Stirling Radioisotope Generator (SRG) is based on two 55-watt electric converters each powered by one GPHS unit. The hot end of the Stirling converter reaches 650°C and heated helium drives a free piston reciprocating in a linear alternator, heat being rejected at the cold end of the engine. The AC is then converted to 55 watts DC. This Stirling engine produces about four times as much electric power from the plutonium fuel than an RTG. Thus each SRG will deliver 100-120 watts of electric power. The SRG has been extensively tested but has not yet flown in space.

Russia has developed RTGs using Po-210, and two of these are still in orbit on 1965 *Cosmos* navigation satellites. But it concentrated on fission reactors for its space power systems (see below).

As well as RTGs, radioactive heater units (RHUs) are used on satellites and spacecraft to keep instruments and batteries warm enough to function efficiently. Their output is only about one watt and they mostly use Pu-238 – typically about 2.7g of it. Some 240 have been used so far by the USA and two are in shut-down Russian Lunar Rovers on the moon. Each of the US Mars Rovers which landed in 2004 uses eight of them to keep the batteries functional.

Both RTGs and RHUs are designed to survive major launch and re-entry accidents intact, as is the SRG.

Fission systems – heat

For power requirements over 100 kWe, fission power systems have a distinct cost advantage over RTGs.

They have been used mainly by Russia. New and more powerful designs are under development.

The US SNAP-10A launched in 1965 was a 45 kWt thermal nuclear fission reactor which supplied 650 watts using a thermoelectric converter. It operated for 43 days, after which it had to be shut down due to a satellite (not reactor) malfunction. It remains in orbit.

The last US space reactor initiative was a joint NASA-Department of Energy-Department of Defense program developing the SP-100 reactor. This was a 2 MWt fast neutron reactor unit and thermoelectric system delivering up to 100 kWe as a multi-use power supply for orbiting missions or as a Lunar/Martian surface power station. The initiative was terminated in the early 1990s after absorbing nearly $1 billion. The reactor used uranium nitride fuel and was lithium-cooled.

Between 1967 and 1988 Russia launched 31 low-powered fission reactors in Radar Ocean Reconnaissance Satellites (RORSATs) on *Cosmos* missions. They utilised thermoelectric converters to produce electricity, as with the RTGs. Romashka reactors were their initial nuclear power source, a fast spectrum graphite reactor with 90%-enriched uranium carbide fuel operating at high temperature. Later reactors, such as on Cosmos-954, which re-entered over Canada in 1978, had U-Mo fuel rods and a layout similar to the US heatpipe reactors described below. It left a trail of radioactive debris spread over Canada's Northwest Territories.

These were followed by the Topaz reactors with thermionic conversion systems, supplying about 5 kWe of electricity for on-board uses. This was a US idea developed during the 1960s in Russia. Topaz-1 was flown in 1987 on two *Cosmos* missions. It was capable of delivering power for 3-5 years for ocean surveillance. Later Topaz were aiming for 40 kWe via an international project undertaken largely in the USA from 1990. Two Topaz-II reactors (without fuel) were sold to the USA in 1992. Budget restrictions in 1993 forced cancellation of the Nuclear Electric Propulsion Space Test Program associated with this.

Development of a small fission surface power system for the Moon and Mars was announced by NASA in 2008. The 40 kWe system could utilise one of two design concepts for power conversion. The first would use two opposed piston engines coupled to alternators that supply 6 kWe each, for a total of 12 kWe of power. The second would develop a closed Brayton cycle engine with a high-speed turbine and compressor coupled to a rotary alternator, also supplying 12 kWe of power.

Fission systems – space propulsion

For spacecraft propulsion, once launched, some experience has been gained with nuclear thermal rocket (NTR) propulsion systems. Nuclear fission heats a hydrogen propellant which has been stored as liquid in cooled tanks. The hot gas (about 2500°C) is expelled through a nozzle to give thrust (which may be augmented by injection of liquid oxygen into the supersonic hydrogen exhaust). This is more efficient than chemical reactions. Bimodal versions will run electrical systems on board a spacecraft, including powerful radars, as well as providing propulsion. Compared with nuclear electric plasma systems (see below), these have much more thrust for shorter periods and can be used for launches and landings.

However, attention is now turning to nuclear electric plasma systems, where nuclear reactors are a heat source for electric ion drives expelling plasma out of a nozzle to propel spacecraft already in space. Superconducting magnetic cells ionise hydrogen or xenon, heat it to extremely high temperatures (millions °C), accelerate it and expel it at very high velocity (*e.g.* 30 km/s) to provide thrust.

Heatpipe power system (HPS) reactors are compact fast reactors supplying up to 100 kWe for about ten years to power a spacecraft or planetary surface vehicle. They have been developed since 1994 at the Los Alamos National Laboratory as a robust and low technical risk system with an emphasis on high reliability and safety. They employ heatpipes to transfer energy from the reactor core to make electricity using Stirling or Brayton cycle converters. Energy from fission is conducted from the fuel pins to the heatpipes filled with sodium vapour which carry it to the heat exchangers and thence in hot gas to the power conversion systems to make electricity. The gas is 72% helium and 28% xenon. The reactor itself contains a number of heatpipe modules with the fuel. Each module has its central heatpipe with rhenium-clad fuel sleeves arranged around it. They contain 97% enriched uranium nitride fuel, all within the cladding of the module. The modules form a compact hexagonal core.

The SAFE-400 (Safe Affordable Fission Engine) space fission reactor is a 400 kWt HPS supplying 100 kWe to power a space vehicle using two Brayton power systems – gas turbines driven directly by the hot gas from the reactor. A smaller version of this kind of reactor is the HOMER-15 – the Heatpipe-Operated Mars Exploration Reactor. It is a 15 kW thermal unit similar to the larger SAFE model, and stands 2.4 metres tall including its heat exchanger and 3 kWe Stirling engine (see section on *Radioisotope systems* above). Total mass of reactor system is 214 kg, and diameter is 41 cm.

Russia is aiming to design a space nuclear propulsion and generation installation in the megawatt power range by 2017 for the exploration of the Moon and Mars. Russia's Energia space corporation started work in 2011 on standardized space modules with nuclear-powered propulsion systems, initially involving 150 to 500 kilowatt systems. The basic engineering design is expected in 2012, with life-service tests planned for 2018. The first launches are envisaged for about 2020.

In 2002, NASA launched Project Prometheus, whose purpose was to enable a major step change in the capability of space missions. Nuclear-powered space travel would be much faster than is now possible, and this would enable manned missions to Mars. One part of Prometheus, with substantial involvement by the US Department of Energy, was to develop the Multi-Mission RTG and the Stirling Radioisotope Generator described in the *Radioisotope systems* section above.

A more radical objective of Prometheus was to produce a space fission reactor system such as those described above for both power and propulsion that would be safe to launch and which would operate for many years with much greater power than RTGs. However, the program has been put on hold.

See also WNA information paper *Nuclear Reactors for Space*.

7.7 RESEARCH REACTORS, MAKING RADIOISOTOPES

Many of the world's nuclear reactors are used for research and training, materials testing, doping silicon for semiconductors, or the production of radioisotopes for medicine and industry. These are much smaller than power reactors or those propelling ships, and many are on university campuses. There are about 240 such reactors operating, in 56 countries – 92 of them in developing countries.

Research reactors comprise a wide range of civil and commercial nuclear reactors, which are generally not used for power generation. The primary purpose of research reactors is to provide a neutron source, both for research and other purposes. Their output (neutron beams) can have different characteristics depending on use. The reactors are small relative to power reactors, whose primary function is to produce heat to make electricity. Their power is designated in megawatts, or kilowatts, thermal (MWt or kWt), but here we will use simply MW (or kW). Most range up to 100 MW, compared with 3000 MW (*i.e.* about 1000 MWe) for a typical power reactor. In fact the total power of the world's 240 or so research reactors is little over 3000 MW.

Research reactors are simpler than power reactors and operate at lower temperatures. They need far less fuel, and far fewer fission products build up as the fuel is used. On the other hand, their fuel requires more highly enriched uranium, typically up to 20% U-235, although many older ones used what is defined as highly-enriched uranium (over 20% U-235). They also have a very high power density in the core, which (though relatively small) requires special design features. Like power reactors, the core needs cooling, and usually a moderator is required to slow down the neutrons and enhance fission. As neutron production is their main function, most research reactors also need a reflector to reduce neutron loss from the core.

Types of research reactor

There is a wider array of designs in use for research reactors than for power reactors, where 80% of the world's plants are of just two similar types. They also have different operating modes, producing energy which may be steady or pulsed.

A common design (about 67 units) is the pool type reactor, where the core is a cluster of fuel elements sitting in a large pool of water. Among the fuel elements are control rods and empty channels for experimental materials. Each element comprises several (*e.g.* 18) curved aluminium-clad fuel plates in a vertical box. The water both moderates and cools the reactor, and graphite or beryllium is generally used for the reflector. Apertures to access the neutron beams are set in the wall of the pool. Tank type research reactors (about 32 units) are similar, except that cooling is more active.

The TRIGA reactor is another common design (about 40 units). The core consists of 60-100 cylindrical fuel elements about 36 mm diameter with aluminium cladding enclosing a mixture of uranium fuel and zirconium hydride (as moderator). It sits in a pool of water and generally uses graphite or beryllium as a reflector. This kind of reactor can safely be pulsed to very high power levels (*e.g.* 25,000 MW) for fractions of a second. Its fuel gives the TRIGA a

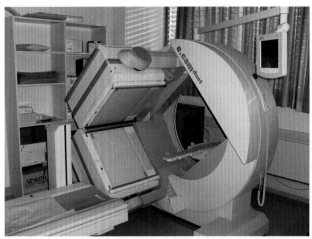

Gamma camera for Positron Emission Tomography in diagnostic nuclear medicine

very strong negative temperature coefficient, and the rapid increase in power is quickly cut short by a negative reactivity effect of the moderator.

Other designs are moderated by heavy water (12 units) or graphite. A few are fast reactors, which require no moderator and can use a mixture of uranium and plutonium as fuel. Homogenous type reactors have a core comprising a solution of uranium salts as a liquid, contained in a tank about 300 mm diameter. The simple design made them popular early on, but only five are now operating.

The International Atomic Energy Agency (IAEA) lists several categories of broadly-classified research reactors. They include critical assemblies – usually zero power (60), test reactors (23), training facilities (37), and two prototypes. But most (160) are largely for research, although some may also produce radioisotopes. As expensive scientific facilities, they tend to be multi-purpose, and many have been operating for more than 30 years.

Russia has most research reactors (62), followed by the USA (54), Japan (18), France (15), Germany (14) and China (13). Many small and developing countries also have research reactors, including Bangladesh, Algeria, Colombia, Ghana, Nigeria, Jamaica, Libya, Thailand and Vietnam. About 20 more reactors are planned or under construction, and over 400 have been shut down or decommissioned, about half of these in USA. Many research reactors were built in the 1960s and 1970s. The peak number operating was in 1975, with 373 in 55 countries.

Uses

Research reactors have a wide range of uses, including analysis and testing of materials, and production of radioisotopes. Their capabilities are applied in many fields, within the nuclear industry as well as in fusion research, environmental science, advanced materials development, drug design and nuclear medicine.

Nuclear medicine uses radiation to provide information about the functioning of specific organs or to treat disease. In most cases, the information is used by physicians to make a quick, accurate diagnosis. The thyroid, bones, heart, liver and many other organs can be easily imaged, and disorders in their function revealed. In some cases radiation can be used to treat diseased organs, or tumours. Over 10,000 hospitals worldwide use radioisotopes in medicine, and about 90% of the procedures are for diagnosis. The most common radioisotope used in diagnosis is technetium-99, with some 30 million procedures per year, accounting for 80% of all nuclear medicine procedures worldwide. Commercial Tc-99 production depends on research reactors.

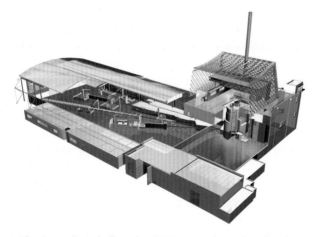

The Bragg beamhall at the OPAL research reactor, showing several beamlines used for different research purposes. (ANSTO)

Neutron beams are uniquely suited to studying the structure and dynamics of materials at the atomic level. For example, 'neutron scattering' is used to examine samples under different conditions such as variations in vacuum pressure, high temperature, low temperature, and magnetic field.

Using neutron activation analysis, it is possible to determine accurately the composition of minute quantities of material. Atoms in a sample are made radioactive by exposure to neutrons in a reactor. The characteristic radiation each element emits can then be detected.

Neutron activation is also used to produce radioisotopes, widely used in industry and medicine, by bombarding particular elements with neutrons. For example, yttrium-90 microspheres to treat liver cancer are produced by bombarding yttrium-89 with neutrons. The most widely

used isotope in nuclear medicine is technetium-99, a decay product of molybdenum-99. Mo-99 is produced by irradiating uranium foil enriched in U-235 with neutrons, and then separating the molybdenum from the other fission products in a hot cell. Most Tc-99 production has been using highly-enriched uranium (HEU) targets, but increasingly low-enriched uranium (LEU) is favoured for non-proliferation reasons, and HEU is being phased out.

Research reactors can also be used for industrial processing. Neutron transmutation doping makes silicon crystals more electrically conductive for use in electronic components.

In test reactors, materials are subject to intense neutron irradiation to study changes. For instance, some steels become brittle, and alloys which resist embrittlement must be used in nuclear reactors.

Like power reactors, research reactors are covered by IAEA safety inspections and safeguards, because some have potential for making nuclear weapons. India's 1974 atomic explosion was the result of plutonium production in a large, but internationally unsupervised, research reactor of a type which is well recognised as having that potential and which closed at the end of 2010.

Fuels

Fuel assemblies are typically plates or cylinders of uranium-aluminium alloy (U-Al) clad with pure aluminium. They are quite different from the ceramic UO_2 pellets enclosed in zircaloy cladding usually used in power reactors. Only a few kilograms of uranium is needed to fuel a research reactor, albeit more highly enriched, compared with perhaps a hundred tonnes in a power reactor.

Some research reactors operate with highly-enriched uranium fuel, but international efforts are under way to substitute it with low-enriched fuel. Highly-enriched uranium (HEU – enriched to more than 20% U-235) allows more compact cores, with high neutron fluxes and also longer times between refuelling. Therefore many reactors up to the 1970s used it.

Since the early 1970s security concerns have grown, especially since many research reactors are located at universities and other civilian locations with much lower security than military weapons establishments where larger quantities of HEU exist. Since 1978 only one reactor, the FRM-II at Garching in Germany, has been built with HEU fuel, while over 20 have been commissioned on low-enriched (LEU) fuel in 16 countries.

The 1980 UN-sponsored International Nuclear Fuel Cycle Evaluation conference concluded that to guard against weapons proliferation from the HEU fuels then commonly used in research reactors, enrichment levels should be reduced to no more than 20% U-235. This followed a similar initiative by the USA in 1978 when its program for Reduced Enrichment for Research and Test Reactors (RERTR) was launched.

Most research reactors using HEU fuel were supplied by the USA and Russia, hence efforts to deal with the problem are largely their initiative. The RERTR program concentrates on reactors over 1 MW which have significant fuel requirements.

These programs have led to the development and qualification of new, high density LEU fuels. The original HEU fuel density was about 1.3-1.7 g/cm^3 uranium. Lowering the enrichment meant that the density had to be increased. Initially this was to 2.3-3.2 g/cm^3 with existing U-Al fuel types.

To September 2009, 67 research reactors (17 in the USA) had been converted to low-enriched uranium silicide fuel or shut down, including major reactors in Ukraine, Uzbekistan and South Africa. Another 34 are convertible using present fuels. A further 28, mostly Russian designs but including two US university reactors, need higher-density fuels not yet available. The goal is to convert or shut 129 reactors by 2018. However, no Russian research reactor has yet been converted to LEU, and the Russian effort has been focused on its reactors in other countries. Russia is now looking at the feasibility of converting six domestic reactors, while others will require high-density fuels.

The first generation of new LEU fuels used uranium silicide dispersed in aluminium (U_3Si_2-Al), at 4.8 g/cm^3. There have been successful tests with denser U_3Si-Al fuel plates up to 6.1 g/cm^3, but US development of these silicide fuels faltered. The presence of silicon makes reprocessing more difficult.

An international effort is under way to develop, qualify and license a high density fuel based on U-Mo alloy dispersed in aluminium, with a density of 6-8g/cm^3. This is to provide fuels which can extend the use of LEU to those reactors requiring higher densities than available in silicide dispersions and to provide a fuel that can be more easily reprocessed than the silicide type.

In a further stage of U-Mo fuel development, which has become the main priority, U-Mo fuel in a monolithic form is

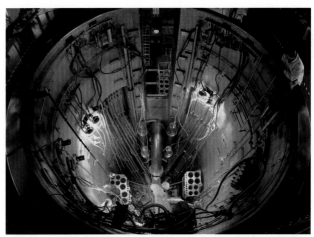

Inside the OPAL research reactor. The core is in the centre at the bottom of the pool. (ANSTO)

being tested, instead of a dispersion of U-Mo in aluminium. The uranium density is 15.6 g/cm^3 and this would enable every research reactor in the world to convert from HEU to LEU fuel without loss of performance. The target date for availability of this was extended to 2013, but is in doubt.

Used research reactor fuel

U-Al fuels can be reprocessed by Areva in France, and U-Mo fuels may also be reprocessed there. U-Si and TRIGA fuels are not readily reprocessed in conventional facilities. However, at least one commercial operator has confirmed that U-Si fuels may be reprocessed in existing plants if diluted with appropriate quantities of other fuels, such as U-Al.

To answer concerns about interim storage of used research fuel around the world, the USA launched a program to take back US-origin spent fuel for disposal and nearly half a tonne of U-235 from such HEU fuel has so far been returned.

Disposal of highly-enriched or even 20% enriched fuel needs to address problems of criticality. This can be achieved by the use of neutron absorbers, or by diluting or dispersing the used fuel in some way within a repository.

In Russia, a parallel trilateral program involving the IAEA and the USA moved two tonnes of HEU and 2.5 tonnes of LEU used fuel from ten countries to the Mayak reprocessing complex near Chelyabinsk over the ten years to 2012. Seventeen countries have Soviet-supplied research reactors, and there are 25 such reactors outside Russia, 15 of them still operational. Since Libya joined the program in 2004, only North Korea objects to it.

See also WNA information paper *Research Reactors.*

8. Environment, health and safety

Production of electricity from any form of primary energy has some environmental effect. These and the health consequences of electricity generation from any source need to be considered. They are known as external costs – those which are quantifiable but do not appear in the utility's accounts. Hence they are not passed on to the consumer, but are borne by society at large. They include particularly the effects of air pollution on human health, crop yields and buildings, as well as occupational disease and accidents. In principle, external costs include the effects on ecosystems and the impact of global warming, though these are much harder to quantify. Carbon taxes or emission trading schemes will bring a carbon dioxide external cost to account in some measure so that it is internalised and passed on to electricity customers in future.

> **The need for clean electricity generation has never been more evident, nor more popularly supported.**

A balanced assessment of nuclear power requires comparison of its environmental effects with those of the principal alternative, coal-fired electricity generation, as well as with other options.

8.1 GREENHOUSE GAS EMISSIONS

The greenhouse effect refers to how certain gases in the Earth's atmosphere allow short-wave radiation in, but trap long-wave thermal radiation emitted from the Earth's surface. This is what keeps the Earth habitable – it would otherwise be too cold at night. However, a build-up of greenhouse gases, notably carbon dioxide (CO_2), appears to be causing a warming of the climate in many parts of the world. If this is continued it will cause changes in weather patterns, and other effects. Much of the greenhouse effect is due to carbon dioxide[1], more than one-third of which comes from power generation.

While our understanding of relevant processes is advancing, we do not know how much CO_2 the environment can absorb, nor exactly how long-term global CO_2 balance is maintained. However, scientists are increasingly concerned about the steady worldwide build-up of CO_2 levels in the atmosphere, and some major political initiatives reflect this. The CO_2 build-up is occurring as the world's carbon-based fossil fuels are being burned and rapidly converted to atmospheric CO_2 e.g. in motor vehicles, domestic and industrial furnaces, and electric power generation. Progressive clearing of the world's forests also contributes to the greenhouse effect by diminishing the removal of atmospheric CO_2 by photosynthesis. These facts are aligned with some evidence that the globe is warming. Debate is centred on the extent to which this might be caused by human activity, and therefore needing to be countered by human action.

As early as 1977, a US National Academy of Sciences report concluded that "the primary limiting factor on energy production from fossil fuels over the next few centuries may turn out to be the climatic effects of the release of carbon dioxide", though the precise effects were uncertain. Today the apparent climate effect of CO_2 has become a widespread concern, and has a major influence on decisions about how electricity should be generated. In particular, it is a significant factor in the comparison of coal and nuclear power for producing electricity.

> **Every 22 tonnes of uranium used in a light water reactor saves about one million tonnes of CO_2 relative to coal.**

Worldwide emissions of CO_2 from burning fossil fuels total about 30 billion tonnes per year. About 40% of this is from coal and about 43% from oil. Every 1000 MWe power station running on black coal produces CO_2 emissions of about 7 million tonnes per year. If brown

---

[1] CO_2 constitutes only 0.039% (390 ppm) of the atmosphere. An increase from 280 to 390 ppm has already occurred since the beginning of the Industrial Revolution.

Figure 23. Greenhouse gas emissions from electricity production

Figure 23. Greenhouse gas emissions from electricity production

Legend:
- Indirect, from life-cycle
- Direct emissions from burning
- *Twin bars indicate range*

Y-axis: Grams CO_2 equivalent /kWh

Data values by category:
- Coal: 289 / 1017 (left bar); 176 / 790 (right bar)
- Gas: 113 / 575 (left bar); 77 / 362 (right bar)
- Hydro: 236 (left bar); 4 (right bar)
- Solar PV: 280 (left bar); 100 (right bar)
- Wind: 48 (left bar); 10 (right bar)
- Nuclear: 21 (left bar); 9 (right bar)

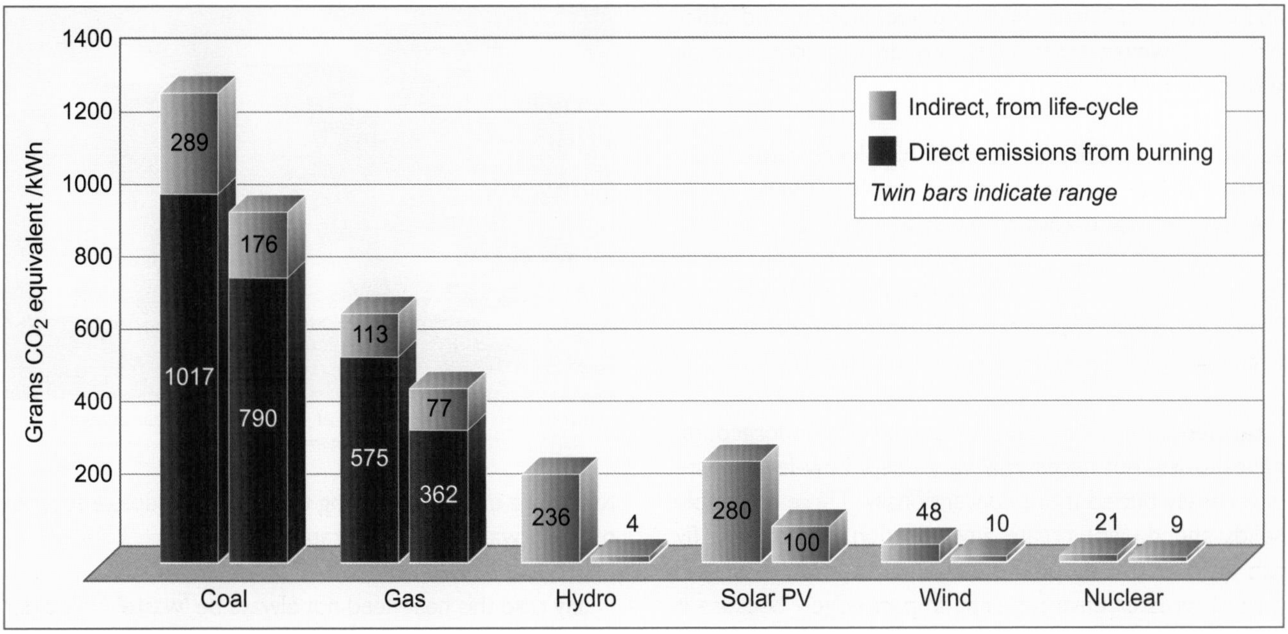

Source: IAEA 2000

coal is used, the amount is about 9 million tonnes. Nuclear fission does not produce any CO_2. Emissions from other parts of the fuel cycle (e.g. uranium mining and enrichment) amount to about 2% of the amount from using coal, and some audited figures show considerably less than this[2].

There is now widespread agreement that we need resource strategies which will minimise CO_2 build-up. Electricity generation contributes about 9.5 billion tonnes of CO_2 per year, so there is plenty of scope for reducing that. In respect to base-load electricity generation, increased use of uranium as a fuel is the most obvious such strategy, utilising proven technology on the scale required (see Figure 23).

There are proposals for capturing the carbon dioxide emissions from burning fossil fuels in large plants such as power stations and then injecting them deep underground – so-called carbon capture and storage (CCS). Though geological disposal of CO_2 has been demonstrated, the effective capture of CO_2 from power stations has not. With much R&D funding now being applied to CCS it is likely that it will become technically practical within a decade, but at considerable extra cost for power generation[3]. The UN Intergovernmental Panel on Climate Change (IPCC) estimated that it will increase the fuel

needs of a power station by at least 25%, and likely double the price of electricity.

8.2 OTHER ENVIRONMENTAL EFFECTS

At a uranium mine, ordinary operating procedures normally ensure that there is no significant water or air pollution. The environmental effect of coal mining today is also small except that more extensive areas will require subsequent rehabilitation, and in certain areas acid mine drainage due to oxidation of sulfides in the rock can be a problem. The effects of uranium mining are discussed more fully in Section 5.1.

Radioactivity
Small amounts of radioactive materials are released to the atmosphere from both coal-fired and nuclear power stations. In the case of coal combustion, small quantities of uranium and thorium present in the coal cause the ash to be radioactive, the level varying considerably. Nuclear power stations and reprocessing plants release small quantities of radioactive gases (e.g. krypton-85 and xenon-133) and iodine-131, which may be detectable in the environment with sophisticated monitoring and analytical equipment. Steps are being taken to further reduce

---

[2] See WNA information paper on *Energy Balances and CO_2 Implications*.
[3] See WNA information paper on *'Clean Coal' Technologies, Carbon Capture & Sequestration*.

emissions of both fly ash from coal-fired power stations and radionuclides from nuclear power stations and other plants. However, these emissions do not constitute an environmental problem.

As outlined in Chapter 6, solid high-level waste from nuclear power stations is stored for 40-50 years while the radioactivity decays to much less than one percent of its original level. Then it will be finally disposed of deep underground and well away from the biosphere. There has been no pollution from such material and nor is any likely, either short- or very long-term.

Intermediate-level waste is placed in underground repositories, not necessarily very deep. Low-level waste is generally buried more conventionally. These are stable solids and do not create any pollution. Radioactive fly ash from coal-fired power stations has in the past had a much greater environmental impact largely because it was not perceived as a problem and appropriate controls were not implemented. Some has been used in building materials. Today most fly ash is removed from stack gases and with bottom ash is buried where seepage and run-off can be controlled.

Waste heat

Waste heat produced due to the intrinsic inefficiency of energy conversion, and hence as a by-product of power generation, is much the same whether coal or uranium is the primary fuel. The thermal efficiency of coal-fired power stations ranges up to a possible 40%, with newer ones typically giving better than 35%. That of nuclear stations in service mostly ranges from 29 - 38% with the common light water reactor today giving about 34%. Waste heat from coal-fired power stations is released both through the condenser circuit and the exhaust stack, that from nuclear stations is all through the condenser circuit, either to large water bodies or via cooling towers.[4]

There is no reason for preferring one fuel over the other on account of waste heat and consequent water requirements for cooling. This is the case whether power station cooling is by water from a stream or estuary, or using atmospheric cooling towers which evaporate water. However, it is noteworthy that whereas coal-fired power plants tend to be located near a source of coal, nuclear plants can be sited according to cooling requirements, and can more readily make use of lake or sea water for direct cooling. Hence they are less likely

Natural draft cooling towers at Mochovce nuclear power plant, Slovakia (Slovenské elektrárne)

to require expensive cooling towers or to deplete supplies of fresh water for evaporative cooling.

In any case this heat need not always be 'waste'. In colder climates, district heating and agricultural uses are increasingly found. These decrease the extent to which local fogs result from the release of heat to the environment. In dryer climates, the rejected heat can be used for desalination to provide potable water.

Sulfur dioxide

The main environmental matter relevant to power generation is the production of carbon dioxide (CO_2) and sulfur dioxide (SO_2) as a result of coal-fired electricity generation. When coal of say 2.5% sulfur is used to produce the electricity for one person in an industrialised country for one year, then about 9 tonnes of CO_2 and 120 kilograms of SO_2 are produced (see Figure 7 on page 20).

Sulfur dioxide emissions arise from the combustion of fossil fuels containing sulfur, as many do. Released in large quantities to the atmosphere it can cause (sulfuric) 'acid rains' in areas downwind. In the northern hemisphere many millions of tonnes of SO_2 are released annually from electricity generation, though such pollution has been dramatically reduced from earlier levels. The acid rain (rainwater having a pH of 4 and lower) in northeastern USA and Scandinavia causes ecological changes and economic loss. In the UK and the USA, electric power utilities at first sought to minimise this by increasing their use of natural gas.

It is possible to remove a lot of the SO_2 from coal stack gases using flue gas desulfurization equipment, but the

---

[4] See WNA information paper on *Cooling Power Plants.*

cost is considerable. Power utilities have spent many billions of dollars on this. In the USA, with half its electricity from coal, SO_2 emissions declined from 15.7 million tonnes in 1980 to 4.6 Mt in 2010. On the other hand, between 1980 and 1986, SO_2 emissions in France were halved simply by replacing fossil fuel power stations with nuclear ones. At the same time, electricity production increased 40% and France became a significant exporter of electricity.

Nitrogen oxides

Oxides of nitrogen (NO_x) from fossil fuel power stations operating at high temperatures are also an environmental problem, regardless of fuel source. If high levels of hydrocarbons[5] are present in the air, nitrogen oxides react with them in sunlight to form photochemical smog. Moreover, in the upper atmosphere, oxides of nitrogen can deplete the Earth's ozone layer, increasing the amount of ultra-violet light reaching the Earth's surface.

8.3 HEALTH EFFECTS OF POWER GENERATION

Here the emphasis is on comparing nuclear power with coal-fired power plants for electricity. Both occupational and environmental health effects on people are considered, along with other risks.

Occupational health and safety

Traditionally, occupational health risks have been measured in terms of immediate accident, especially fatality, rates. However, today, and particularly in relation to nuclear power, there is an increased emphasis on less obvious or delayed effects of exposure to cancer-inducing substances and radiation.

Many occupational accident statistics have been generated over the last 55 years of nuclear reactor operations in the USA and the UK. These can be compared with those from coal-fired electricity generation. All show that nuclear is distinctly the safer means of electric power generation in this respect. Two simple sets of figures are quoted in Tables 14 & 15. A major reason for coal showing up unfavourably is the huge amount of it which must be mined and transported to supply even a single large power station – some 15,000 times as much coal as uranium. Mining and multiple handling of so much material of any kind involves hazards, and these are reflected in the statistics.

Table 14. Comparison of accident statistics in primary energy production

Fuel	Immediate fatalities 1970-92	Who?	Normalised to deaths per TWyr* electricity
Coal	6400	Workers	342
Natural gas	1200	Workers & public	85
Hydro	4000	Public	883
Nuclear	30	Workers	8

* Basis: per million MWe (i.e. about three times world nuclear power capacity) operating for one year, not including plant construction, based on historic data – which is unlikely to represent current safety levels in any of the industries concerned. The data in this column was published in 2001 but is consistent with earlier figures, where it was pointed out that the coal total would be about ten times greater if accidents with less than five fatalities were included.
(Electricity generation accounts for about 40% of total primary energy.)
Sources: Ball, Roberts & Simpson, Research Report #20, Centre for Environmental & Risk Management, University of East Anglia, 1994; Hirschberg et al, Paul Scherrer Institut, 1996 in: IAEA, Sustainable Development and Nuclear Power, 1997; Severe Accidents in the Energy Sector, Paul Scherrer Institut, 2001.

Figure 24. Deaths from energy-related accidents per unit of electricity

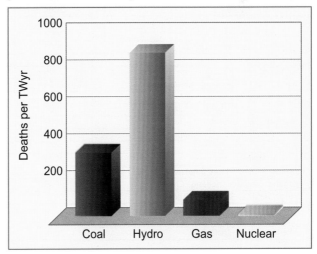

Source: Paul Scherrer Institut 1998, considering 1943 accidents with more than five fatalities.
One TWyr is the amount of electricity used by the world in about five months.

---

[5] Mostly from vehicles.

Table 15. The hazards of using energy: some energy-related accidents since 1977

Place	Year	Number killed	Comments
Machhu II, India	1979	2500	Hydroelectric dam failure
Hirakud, India	1980	1000	Hydroelectric dam failure
Ortuella, Spain	1980	70	Gas explosion
Donbass, Ukraine	1980	68	Coal mine methane explosion
Israel	1982	89	Gas explosion
Guavio, Colombia	1983	160	Hydroelectric dam failure
Nile River, Egypt	1983	317	LPG explosion
Cubatao, Brazil	1984	508	Oil fire
Mexico City	1984	498	LPG explosion
Tbilisi, Russia	1984	100	Gas explosion
Northern Taiwan	1984	314	3 coal mine accidents
Chernobyl, Ukraine	1986	49+	**Nuclear reactor accident**
Piper Alpha, North Sea	1988	167	Explosion of offshore oil platform
Asha-ufa, Siberia	1989	600	LPG pipeline leak and fire
Dobrnja, Yugoslavia	1990	178	Coal mine
Hongton, Shaanxi, China	1991	147	Coal mine methane explosion
Belci, Romania	1991	116	Hydroelectric dam failure
Kozlu, Turkey	1992	272	Coal mine methane explosion
Cuenca, Equador	1993	200	Coal mine
Durunkha, Egypt	1994	580	Fuel depot hit by lightning
Seoul, S. Korea	1994	500	Oil fire
Minanao, Philippines	1994	90	Coal mine
Dhanbad, India	1995	70	Coal mine
Taegu, S. Korea	1995	100	Oil & gas explosion
Spitsbergen, Russia	1996	141	Coal mine
Henan, China	1996	84	Coal mine methane explosion
Datong, China	1996	114	Coal mine methane explosion
Henan, China	1997	89	Coal mine methane explosion
Fushun, China	1997	68	Coal mine methane explosion
Kuzbass, Siberia	1997	67	Coal mine methane explosion
Huainan, China	1997	89	Coal mine methane explosion
Donbass, Ukraine	1998	63	Coal mine methane explosion
Liaoning, China	1998	71	Coal mine methane explosion
Warri, Nigeria	1998	500+	Oil pipeline leak and fire
Donbass, Ukraine	1999	50+	Coal mine methane explosion
Donbass, Ukraine	2000	80	Coal mine methane explosion
Muchonggou, Guizhou, China	2000	162	Coal mine methane explosion
Jixi, China	2002	124	Coal mine methane explosion
Gaoqiao, SW China	2003	234	Gas well blowout with H_2S
Kuzbass, Russia	2004	47	Coal mine methane explosion
Donbass, Ukraine	2004	36	Coal mine methane explosion
Henan, China	2004	148	Coal mine methane explosion
Chenjiashan, Shaanxi, China	2004	166	Coal mine methane explosion
Sunjiawan, Liaoning, China	2005	215	Coal mine methane explosion
Fukang, Xinjiang, China	2005	83	Coal mine methane explosion
Xingning, Guangdong, China	2005	102	Coal mine flooding
Dongfeng, Heilongjiang, China	2005	164	Coal mine methane explosion
Bhatdih, Jharkhand, India	2006	54	Coal mine methane explosion
Ulyanovskaya, Kuzbass, Russia	2007	150	Coal mine methane or dust explosion
Zhangzhuang, Shandong, China	2007	181	Coal mine flooding
Zasyadko, Donetsk, E. Ukraine	2007	101-111	Coal mine methane explosion
Linfen City, Shanxi, China	2007	105	Coal mine methane explosion
Tunlan, Shanxi, China	2009	78	Coal mine methane explosion
Sayano-Shushenskaya, Khakassia, Russia	2009	75	Hydropower turbine disintegration
Hegang City, Heilongjiang, China	2009	108	Coal mine methane explosion
Sanga/Sangha, Bukavu, Congo	2010	235	Petrol tanker accident and fire
Deepwater Horizon, USA	2010	11	Oil well blowout, over 4 million barrels of oil caused massive pollution in Gulf of Mexico
Pike River, New Zealand	2010	29	Coal mine methane explosion

LPG and oil accidents with less than 300 fatalities, and coal mine accidents with less than 100 fatalities are generally not shown unless recent. Coal mining deaths range from 0.009 per million tonnes of coal mined in Australia through 0.034 in USA to more than one in China and Ukraine. China's death rate in 2008 fell to 1.182 per million tonnes of coal mined, compared with 1.485 in 2007, and 3.08 in 2005. China's total death toll from coal mining to 2008 averaged well over 4000 per year – official figures give 5300 in 2000, 5670 in 2001, 6995 in 2002, about 6400 in 2003, 6027 in 2004, about 6000 in 2005, 4746 in 2006, 3786 in 2007 and 3210 in 2008. These data omit the small illegal collieries. However, the picture is improving: in the 1950s the annual death toll in coal mines was 70,000, in the 1980s it was 40,000 and 1990s it was 10,000. Ukraine's coal mine death toll has been over 200 per year (e.g. 1999: 274; 1998: 360; 1995: 339; 1992: 459). Sources: contemporary media reports; Paul Scherrer Inst, 1998 report; International Federation of Chemical, Energy, Mine and General Workers' Unions (ICEM) website (www.icem.org)

Health risks in uranium mining are largely discussed in Section 5.1. In the 1950s, exposure of miners to radon gas led to an increased incidence of lung cancer. For over 40 years, however, exposure to high levels of radon has no longer been a feature of uranium (or other) mines. Today, the presence of some radon around a uranium mining operation and some dust bearing radioactive decay products, as well as the hazards of inhaled coal dust in a coal mine, are well understood. In both cases, using the best current practice, the health hazards to miners are very small and certainly less than the risks of industrial accidents.

In other parts of the nuclear fuel cycle, radiation hazards to workers are low, and industrial accidents are few. Certainly nuclear power generation is not completely free of hazards in the occupational sense, but it has been shown to be far safer than other forms of energy conversion, as Table 14 shows, over more than 20 years.

The occurrence of cancer is not uniform across the world population, and because of local differences it is not easy to see whether or not there is any association between low occupational radiation doses and possible excess cancers. This question has been studied closely in a number of areas and work is continuing. So far no clear evidence has emerged to indicate that cancers are more frequent in radiation workers than in other people of similar ages in Western countries. Nor, incidentally, are they greater in people exposed to very high natural levels of radiation in certain parts of the world – significantly higher than levels allowed in industry. At the low levels of exposure and dose rates involved in the nuclear industry, the effects are probabilistic rather than measurable, as described in Section 8.4.

Environmental hazards

Environmental (non-occupational) health effects are qualitatively similar to those affecting workers in the industry. Popular concern about ionising radiation initially grew out of the testing of nuclear weapons. Correspondingly, these tests provided the nuclear power industry with a strong awareness of radiation hazards. Fortunately radioactivity is readily measurable and its effects are fairly well understood compared with those of other hazards with delayed effects – including virtually all chemical cancer-inducing substances. Radiation is a weak carcinogen.

The contrast between air quality effects from coal burning for electricity and increased radiation from nuclear power is very marked: a person living next to a nuclear power plant receives less radiation from it than from a few hours flying each year (see Table 16). On the other hand, anyone downwind of a coal-fired power plant can expect it to have an effect on the air quality.

8.4 RADIATION EXPOSURE

Table 16 shows some typical levels and sources of radiation exposure. The contribution from the ground and buildings varies from place to place. In most parts of the world, levels range up to 3 millisieverts per year (mSv/yr). Citizens of Cornwall, UK, receive an average of about 7 mSv/yr. Hundreds of thousands of people in India, Brazil and Sudan receive up to 40 mSv/yr. Several places are known in Iran, India and Europe where natural background radiation gives an annual dose of more than 50 mSv, and at Ramsar in Iran it can give up to 260 mSv. Lifetime doses from natural radiation range up to several thousand millisieverts. However, there is no evidence of increased cancers or other health problems arising from these high natural levels.

Cosmic radiation dose varies with altitude and latitude. Aircrew can receive up to about 5 mSv/yr from their hours in the air, while frequent flyers can score a similar increment. In contrast, UK citizens receive about 0.0003 mSv/yr from nuclear power generation. Appendix 1 gives further background to the topic of radiation and its measurement.

In practice, radiation protection is based on the understanding that small increases over natural levels of exposure are not harmful but should prudently be kept to a minimum. To put this into practice the International Commission for Radiological Protection (ICRP) has established recommended standards of protection based on three basic principles:

- Justification. No practice involving exposure to radiation should be adopted unless it produces a net benefit to those exposed or to society generally.
- Optimisation. Radiation doses and risks should be kept as low as reasonably achievable (ALARA), economic and social factors being taken into account.
- Limitation. The exposure of individuals should be subject to dose or risk limits above which the radiation risk would be deemed unacceptable.

These principles apply to the potential for accidental exposures as well as predictable normal exposures.

Underlying these principles is the application of the 'linear hypothesis' based on the idea that any level of radiation dose, no matter how low, involves the

Table 16. Ionising radiation

The Earth is radioactive, due to the decay of natural long-lived radioisotopes. Radioactive decay results in the release of ionising radiation. As well as the Earth's radioactivity we are naturally subject to cosmic radiation from space. In addition to both these, we collect some radiation doses from artificial sources such as X-rays. We may also collect an increased cosmic radiation dose due to participating in high altitude activities such as flying or skiing. The average adult contains about 13 mg of radioactive potassium-40 in body tissue – we therefore even irradiate one another at close quarters!

The relative importance of these various sources is indicated below. Types of radiation and units for measuring it are outlined in Appendix 1.

	Typical μSv/yr	Range μSv/yr
Natural:		
Terrestrial + house: radon	200	200-100,000
Terrestrial + house: gamma	600	100-1,000
Cosmic	300 at sea level	
	(+20 for every 100m elevation)	0-500
Food, drink & body tissue	400	100-1,000
Total	**1,500** (plus altitude adjustment)	
Artificial:		
From nuclear weapons tests	3	
Medical (X-ray, CT etc. average)	370	up to 75,000
From nuclear energy	0.3	
From coal burning	0.1	
From household appliances	0.4	
Total	**375**	
Behavioural:		
Skiing holiday	8 per week	
Air travel in jet airliner	1.5-5 per hour	up to 5,000 per year

The International Commission for Radiological Protection (ICRP) recommends, in addition to background, the following exposure limits:

For general public	1,000 (*i.e.* 1 mSv/yr)
For nuclear worker	20,000 (*i.e.* 20 mSv/yr) averaged over five consecutive years

Sources: Australian Radiation Protection & Nuclear Safety Agency, Health Protection Agency (UK), Australian Nuclear Science & Technology Organisation

possibility of risk to human health. This assumption enables 'risk factors' derived from studies of high radiation dose to populations (*e.g.* from Japanese atomic bomb survivors) to be used in suggesting the risk to an individual from low doses[6]. However the weight of scientific evidence does not indicate any cancer risk or immediate effects at doses below 50 mSv in a short

time or at about 100 mSv per year. At lower doses and dose rates (up to at least 10 mSv/yr) the evidence suggests that beneficial effects are at least as likely as harmful ones.

Based on the three conservative principles, the ICRP recommends that the additional dose above natural

[6] ICRP Publication 60

background and excluding medical exposure should be limited to prescribed levels. These levels are: 1 mSv/yr for members of the public, and 20 mSv/yr averaged over five years for radiation workers, who are required to work under closely-monitored conditions (see Table 16).

Effects of radiation

The actual level of individual risk at the ICRP-recommended limit for general public exposure is negligible (it is calculated to result in about one fatal cancer per year in a population of 20,000 people) and impossible to confirm directly. In the Chernobyl accident (see Section 8.5), a large number of people were subject to significantly increased radiation exposure, the actual doses being approximately known. In the Fukushima accident, few workers and very few others were subject to radiation exposure at levels of concern. Both accidents will result in a better understanding of the effects, if any, of exposure to various levels of radiation.

At present much of our knowledge about the effect of radiation on people is derived from the survivors of the Hiroshima and Nagasaki bombings in 1945, where the doses received were very brief and also difficult to estimate. Certainly there was a clear increase in certain types of leukaemia and lymphoma and other solid cancers among the survivors. Progressively, there is more information based on exposure with low dose rate, where the body has time to repair damage. The human body has defence mechanisms against damage induced by radiation as well as by chemical carcinogens. These can be stimulated by low levels of exposure, or overwhelmed by very high levels[7]. Multiplying very small dose figures by very large numbers of people is meaningless, a conclusion now supported by the ICRP.

Plutonium is sometimes seen as a particular concern. It can be separated from used fuel by reprocessing, as discussed in Section 6.2. Plutonium has been called the 'most toxic element known' and therefore a hazard that we should do without. However such a claim must be tested by comparing its toxicity with that of other materials familiar to us or around us. If swallowed, plutonium is much less toxic than cyanide or lead arsenate and about twice as toxic as the concentrate of caffeine from coffee. Its main danger comes if inhaled as a fine dust and absorbed through the lungs. This would increase the likelihood of

cancer 15 or more years afterwards. However, as a counterpoint to the folklore about plutonium is the fact that about seven tonnes of it were dispersed in the upper atmosphere by nuclear weapons testing over the 30 years following World War II without identifiable ill effects.

The health effects of exposure both to radiation and to chemical cancer-inducing agents or toxins must be considered in relation to time. We should be concerned not only about the effects on people presently living, but also about the cumulative effects of actions today over many generations. Some radioactive materials which reach the environment decay to safe levels within days, weeks or a few years, while others continue their effect for a long time, as do some chemical cancer-inducing agents and toxins. Certainly this is true of the chemical toxicity of heavy metals such as mercury, cadmium and lead, these of course being a natural part of the human environment anyway, like radiation, but maintaining their toxicity for ever. The essential task for those in government and industry is to prevent excessive amounts of such toxins harming people, now or in the future. Standards are set in the light of research on environmental pathways by which people might ultimately be affected.

About 60 years ago it was discovered that ionising radiation could induce genetic mutations in fruit flies. Intensive study since then has shown that radiation can similarly induce mutations in plants and test animals. However, there is no evidence of genetic damage to humans from radiation – even as a result of the large doses received by atomic bomb survivors in Japan.

In a plant or animal cell, the material (DNA) which carries genetic information necessary to cell development, maintenance and division is the critical concern for radiation. Much of the damage to DNA is repairable, but in a small proportion of cells the DNA is permanently altered. This may result in death of the cell or development of a cancer, or in the case of cells forming gonad tissue, alterations which continue as genetic changes in subsequent generations. Most such mutational changes are deleterious, so that the affected line dies out; very few can be expected to result in improvements.

The levels of radiation allowed for members of the public and for workers in the nuclear industry are such that any

---

[7] Tens of thousands of people in each technically-advanced country work in medical and industrial environments where they may be exposed to radiation above background levels. Accordingly they wear monitoring badges while at work, and their exposure is carefully monitored. The health records of these occupationally exposed groups often show that they have lower rates of mortality from cancer and other causes than the general public and, in some cases, significantly lower rates than other workers who do similar work without being exposed to radiation.

increase in genetic effects due to nuclear power will be imperceptible, if not non-existent. Radiation exposure levels are set so as to prevent tissue damage and minimise the risk of cancer. Experimental evidence in laboratories indicates that cancers are more likely than genetic damage. Some 75,000 children born of parents who survived high radiation doses at Hiroshima and Nagasaki in 1945 have been the subject of intensive examination. This study confirms that no increase in genetic abnormalities in human populations is likely as a result of even quite high doses of radiation. Similarly, no genetic effects are evident as a result of the Chernobyl accident.

Life on Earth commenced and developed when the environment was certainly subject to several times as much radioactivity as it is now, so radiation is not a new phenomenon. If we ensure that there is no dramatic increase in people's general radiation exposure, and levels of exposure are kept well below some of those occurring naturally, we can be confident that genetic damage due to radiation will not become significant.

8.5 REACTOR SAFETY

> **The safety of nuclear power reactors has been amply demonstrated over more than 50 years and almost 15,000 reactor-years of operation.**

There have been sophisticated statistical studies on reactor safety. However, for most people actual performance is more convincing than probability statistics. The situation to date is that there have been only two accidents to commercial reactors where the effects were not substantially contained within the design and structure of the reactor. To this experience one could add another 12,000 reactor years from naval operation, which in the West has had an excellent safety record.

It has long been asserted that nuclear reactor accidents are the epitome of low-probability but high-consequence risks. However, the physics and chemistry of a reactor core, coupled with but not wholly depending on the engineering, mean that the consequences of an accident are likely in fact to be much less severe than those from other industrial and energy sources. Experience bears this out.

Only the Chernobyl disaster in 1986 and the Fukushima accident in 2011 resulted in radiation doses to the public

greater than those resulting from exposure to natural sources. Other incidents (and one 'accident', in 1979) have been completely confined to the plant. The Chernobyl tragedy made it clear why such reactors have never been licensed outside the former Soviet Union. Apart from Chernobyl, no nuclear workers or members of the public have ever died as a result of exposure to radiation due to a commercial nuclear reactor accident or incident. This is remarkable for the first six decades of a complex new technology which is being used in 30 countries, some reactors now operating having been built over 40 years ago.

Most of the serious radiological injuries and deaths that occur each year (2 to 4 deaths and many more exposures above regulatory limits) are the result of large uncontrolled radiation sources, such as abandoned medical or industrial equipment. These have nothing to do with nuclear power generation.

Most accident scenarios in nuclear plants involve primarily a loss of cooling. This may lead to the fuel in the reactor core overheating, melting and releasing fission products. Hence there is provision of emergency core cooling systems on standby. In case these should fail, as at Fukushima due to power loss, further protective barriers come into play. In particular, the reactor core is normally enclosed in structures designed to prevent radioactive releases to the environment. Regulatory requirements today are that the effects of any core-melt accident must be confined to the plant itself, without the need to evacuate nearby residents – which was necessary at Fukushima. About one-third of the capital cost of reactors is normally due to engineering designed to enhance the safety of people – both operators and neighbouring communities – if and when things should go wrong. Table 17 shows the international scale for reporting nuclear accidents or incidents.

The main safety concern has always been over the possibility of an uncontrolled release of radioactive material, leading to contamination and consequent radiation exposure to people nearby. Earlier assumptions were that this would be likely in the event of a major loss of cooling accident which resulted in a core melt. Experience up to the 2011 Fukushima accident suggested otherwise, at least for most reactor designs.

In reactor accidents such as at Windscale (a military facility) in 1957, at Chernobyl in 1986 and Fukushima in 2011, where radioactive materials were not fully contained, the principal health hazard was from the spread of radioactive materials, notably volatile fission products such as iodine-131 and caesium-137. These are biologically active, so that if consumed in food, they tend to stay in organs of the body. Iodine-131 has a half-life of eight days, so is a hazard for around the first month (and apparently gave rise to thyroid cancers after the Chernobyl accident). Caesium is soluble and can be taken into the body, but does not concentrate in any particular organs, and is eventually excreted. Caesium-137 has a half-life of 30 years, so is therefore potentially a long-term contaminant of pastures and crops. In addition to these, there is caesium-134 which has a half-life of about two years. While measures can be taken to limit human uptake of iodine-131, (evacuation of area for several weeks, iodine tablets), radioactive caesium can preclude food production from affected land for a long time. Other radioactive materials in a reactor core have been shown to be less of a problem because they are either not volatile (strontium, transuranic elements) or not biologically active (tellurium-132).

Three Mile Island

Studies of the post-accident situation at Three Mile Island in 1979 demonstrated the effectiveness of containment. There was no uncontrolled release of radioactive material and the total radioactivity release from this accident was small, so the maximum dose to individuals living near the power plant was well below internationally-accepted limits, even though much of the core melted and the reactor was written off. Nevertheless, this accident had a pronounced psychological impact, was a severe blow to the US nuclear industry and had an adverse effect on the growth of nuclear capacity in the USA and beyond. More positively it brought about profound changes in the way reactors are run, and in details of their engineering. In retrospect it was a very valuable stimulus to improvements, and had much the same effect on reactor safety as the Comet airliner crashes of the 1950s did on the safety of pressurised jet aircraft – to everybody's benefit since.

Chernobyl

The 1986 accident at Chernobyl in Ukraine was very serious due to the design of the reactor and also its burning fuel which dispersed radioactive contamination far and wide. It cost the lives of 30 staff and firefighters, mostly from the effects of acute radiation exposure. There have also been 1800 cases of thyroid cancer registered in children, most of which were curable, though about ten have been fatal. No increase in leukaemia and other cancers had shown up in the first two decades following the accident, though the World

Health Organization (WHO) expected some increase in cancers in this period. The death toll from delayed health effects may well climb beyond the ten or so thyroid cancer victims. About 130,000 people received significant radiation doses (i.e. above ICRP limits), and are being closely monitored by WHO. Radioactive pollution drifted across a wide area of Europe and Scandinavia, causing disruption to agricultural production and some exposure (small doses) to a large population.

The accident drew public attention to the lack of an adequate containment structure such as is standard on Western reactors. In addition, the particular reactor (RBMK) design was such that the conditions created by the operators led to an uncontrolled increase in power output from the fission process. Under abnormal conditions, all reactor types may experience power increases, but in other designs these are reliably controlled by the reactor shutdown systems and by the design physics. Light water reactors, in which the coolant serves as moderator, automatically reduce power when the coolant/moderator is lost, and can then be shut down using the control rods.

The Chernobyl accident was caused by a combination of design deficiencies and the particular operating procedures resulting from an absence of an adequate safety culture. With assistance from the West, significant safety

Destroyed No.4 reactor at Chernobyl before the shelter structure was built

Table 17. The International Nuclear Event Scale

For prompt communication of safety significance

Level, descriptor	Off-site impact	On-site impact	Defence-in-depth degradation	Examples
7 Major Accident	*Major Release:* Widespread health and environmental effects			Fukushima Daiichi 1-3, 2011 (fuel damage, radiation release and evacuation) Chernobyl, Ukraine, 1986 (fuel meltdown and fire)
6 Serious Accident	*Significant Release:* Full implementation of local emergency plans			Mayak at Ozersk, Russia, 1957 (reprocessing plant criticality)
5 Accident with Off-Site Risks	*Limited Release:* Partial implementation of local emergency plans, or	Severe core damage to reactor core or to radiological barriers		Windscale, UK, 1957 (military); Three Mile Island, USA, 1979 (fuel melting)
4 Accident Mainly in Installation	*Minor Release:* Public exposure of the order of prescribed limits, or	Significant damage to reactor core or to radiological barriers, worker fatality		Saint-Laurent A1, France, 1969 (fuel rupture) & A2 1980 (graphite overheating). Tokai-mura, Japan, 1999 (criticality in fuel plant for an experimental reactor)
3 Serious Incident	*Very Small Release:* Public exposure at a fraction of prescribed limits, or	Major Contamination. Acute health effects to a worker, or	Near Accident. Loss of Defence-in-Depth provisions – no safety layers remaining	Fukushima Daiichi 4, 2011 (fuel pond overheating); Fukushima Daini 1, 2, 4, 2011 (interruption to cooling); Vandellos, Spain, 1989 (turbine fire); Davis-Besse, USA, 2002 (severe corrosion); Paks, Hungary 2003 (fuel damage)
2 Incident	nil	Significant spread of contamination. Overexposure of worker, or	Incidents with significant failures in safety provisions	
1 Anomaly	nil	nil	Anomaly beyond authorised operating regime	
0	nil	nil	No safety significance	
Below Scale	nil	nil	No safety relevance	

Source: International Atomic Energy Agency

improvements have been made to the ten RBMK reactors still in operation in Russia. Russian reactor design has since been standardised on PWR types, with containment structures, and meeting Western safety standards.

Soon after the accident, the destroyed Chernobyl 4 reactor was enclosed in a large concrete shell. The other three RBMK units on the site initially resumed operation, though they have since shut down, the last at the end of 2000.

The 2005 Chernobyl Forum study (revised version published 2006)[8] involved over 100 scientists from eight specialist UN agencies and the governments of Ukraine, Belarus and Russia. Its conclusions are in line with earlier studies, such as the OECD expert report which concluded that "the Chernobyl accident has not brought to light any new, previously unknown phenomena or safety issues that are not resolved or otherwise covered by current reactor safety programs for commercial

8 A 55-page version of the Chernobyl Forum report is available on the web: http://www.iaea.org/Publications/Booklets/Chernobyl/chernobyl.pdf
This involved several UN agencies — IAEA, WHO, UNDP, UNDP, UNSCEAR etc — and the governments of Russia, Ukraine and Belarus.
See also: http://www-ns.iaea.org/meetings/rw-summaries/chernobyl_forum.htm

ENVIRONMENT, HEALTH AND SAFETY

power reactors in OECD Member countries." A very positive outcome of the accident was the creation of the World Association of Nuclear Operators (WANO) which enables the sharing of expertise and experience across the world.

Fukushima

The 2011 accident at the Fukushima Daiichi plant in Japan was precipitated by a major tsunami which took the lives of about 19,000 people and devastated large areas. It closely followed the biggest earthquake experienced in Japan, which had caused an interruption in power supply to the site, due to landslides. The three operating reactors shut down automatically due to the earthquake, as they were designed to do, and plant's standby generators all started up automatically but were drowned an hour later when the tsunami arrived. The seawater pumps serving all safety-related heat exchangers were disabled, so the ultimate heat sink was lost also. The net effect was that the decay heat in the fuel of the three recently shut down reactors had nowhere to go, and while emergency core cooling worked for a while in two units, in the oldest one the core melted after about 15 hours and subsequently went through the bottom of the pressure vessel. Some core melting occurred in the other two later. The reactors had been sited and built in 1960s to early 1970s.

Due to the very high temperatures in the reactor vessels, the zirconium fuel cladding reacted with steam to form hydrogen, and as this was vented, some collected in the top service floors on units 1, 3 and 4, causing hydrogen explosions which spread debris around. Considerable

Table 18. Serious reactor accidents

Reactor	Date	Immediate deaths	Environmental effect	Follow-up action
NRX, Canada (experimental, 40 MWt)	1952	Nil	Nil	Repaired (new core). Closed 1992
Windscale 1, UK (military plutonium-producing pile)	1957	Nil	Widespread contamination, farms affected (c. 1.5 PBq released)	Entombed (filled with concrete). Being dismantled
SL-1, USA (experimental, military, 3 MWt)	1961	Three operators	Very minor radioactive release	Decommissioned
Fermi 1, USA (experimental breeder, 66 MWe)	1966	Nil	Nil	Repaired and restarted, then closed in 1972
Lucens, Switzerland (experimental, 7.5 MWe)	1969	Nil	Very minor radioactive release	Decommissioned
Browns Ferry 1, USA (commercial, 1080 MWe)	1975	Nil	Nil	Repaired
Three Mile Island 2, USA (commercial, 880 MWe)	1979	Nil	Minor short-term radiation dose (within ICRP limits) to public. Delayed release of 200 TBq of Kr-85	Clean-up program complete, in monitored storage stage of decommissioning
Saint Laurent A2, France (commercial, 450 MWe)	1980	Nil	Minor radiation release (80 GBq)	Repaired. Decommissioned 1992
Chernobyl 4, Ukraine (commercial, 950 MWe)	1986	47 staff & fire-fighters, (32 immediate)	Major radiation release across E. Europe and Scandinavia (14,000 PBq or 5200 PBq I-131 eq.)	Reactor entombed Pending dismantling
Vandellos-1, Spain (commercial, 480 MWe)	1989	Nil	Nil	Decommissioned
Greifswald, E. Germany (commercial, 440 MWe)	1989	Nil	Nil	Decommissioned
Fukushima 1-3, Japan (commercial, 1959 MWe)	2011	Nil	Significant local contamination (770 PBq I-131 eq.)	Decommissioned with unit 4

Serious accidents in military, research and commercial reactors. All except Browns Ferry and Vandellos involved damage to or malfunction of the reactor core. At Browns Ferry a fire damaged control cables and resulted in an 18-month shutdown for repairs; at Vandellos a turbine fire made the 17-year old plant uneconomic to repair.
The well-publicised accident at Tokai Mura, Japan, in 1999 was at a fuel preparation plant for experimental reactors, and killed two workers from radiation exposure. Many other such criticality accidents have occurred, some fatal, and practically all in military facilities prior to 1980.

Photo of (from right to left) Fukushima Daiichi units 1, 2, 3 and 4 on 20 March 2011 (Air Photo Service Co. Ltd., Japan)

radioactive material was released to the atmosphere due to venting the containments after some molten fuel had fallen through the bottom of the reactor pressure vessels, and due to the hydrogen explosions. As cooling failed on the first day, evacuations had progressively been ordered and by the evening of the second day the evacuation zone had been extended to 20 km from the plant. The main radionuclide released was volatile iodine-131, together with caesium-137, and the main releases were on day 4.

Meanwhile over 250 staff on site grappled with restoring power and supplementary cooling, then with treating contaminated water. Some 160 workers received doses of 100-250 mSv over some days, and six received more than this, apparently due to inhaling iodine fumes early on. There were no radiation casualties (acute radiation syndrome) as at Chernobyl.

In summary, the cause of the Fukushima accident was the loss of all plant equipment needed to support core cooling because of the effect of the great tsunami, well in excess of what was considered reasonably possible by the plant operator or the regulator. In particular:

- Siting the reactors too close to sea level, given historic evidence of high tsunamis.
- Having the back-up diesel generators, switchgear, batteries, pumps and external heat exchangers vulnerable to inundation and damage.
- Having containment vent systems which were not properly operable without power, and not having passive hydrogen recombiners in reactor containments.

See also: WNA information papers on *Chernobyl Accident, Three Mile Island Accident* and *Fukushima Accident*.

Other reactor accidents

There have been a number of accidents in experimental and military reactors, including a number of melted cores, but none of these has resulted in loss of life outside the actual plant, or long-term environmental contamination. Table 18 of serious reactor accidents includes those in which fatalities have occurred, together with the most serious commercial plant accidents. The list probably corresponds to incidents rating Level 4 or higher on today's International Nuclear Event Scale (see Table 17). It should be emphasised that a commercial-type reactor simply cannot under any circumstances explode like a nuclear bomb.

See also WNA information paper on *Cooperation in the Nuclear Power Industry*.

Terrorism

Since the World Trade Center and Pentagon attacks in the USA in 2001 there has been concern about the consequences of a large aircraft being used to attack a nuclear facility with the purpose of releasing radioactive materials. Various studies on this scenario show that nuclear reactors would be more resistant to these than virtually any other civil installations.

A thorough study undertaken by the Electric Power Research Institute in 2002 concluded that US reactor structures "are robust and (would) protect the fuel from impacts of large commercial aircraft." The analyses used a fully-fuelled Boeing 767-400 of over 200 tonnes as the basis, at 560 km/h – the maximum speed for precision flying near the ground. The wingspan of this aircraft is greater than the diameter of reactor containment buildings and the 4.3 tonne engines are 15 metres apart. Hence analyses focused on single engine direct impact on the centreline and on the impact of the entire aircraft if the fuselage hit the centreline (in which case the engines would ricochet off the sides). In each case no part of the aircraft or its fuel would penetrate the containment.

Looking at the storage pools holding used fuel, similar analyses showed no breach. Dry storage and transport casks retained their integrity, so that there would be no release of radionuclides to the environment from these.

Switzerland's Nuclear Safety Inspectorate studied a similar scenario and reported in 2003 that the danger of any radiation release from such a crash would be low for the older plants and extremely low for the newer ones.

Similarly, the massive structures mean that any terrorist attack even inside a plant (which are well defended) would not result in any significant radioactive releases.

The conservative design criteria which caused most power reactors to be shrouded by massive containment structures has provided peace of mind in this context.

Safety comparisons

Coal-fired power generation has chronic, rather than acute, safety implications for public health. It also has profound safety implications for the mining of coal, with thousands of workers killed each year in coal mines.

Hydro power generation has a record of few but very major events causing thousands of deaths. In 1975 when the Banqiao, Shimantan & other dams collapsed in Henan, China, at least 30,000 people were killed immediately and some 230,000 lives lost overall, with 18 GWe also lost. In 1979 and 1980 in India some 3500 were killed by two hydro-electric dam failures, and in 2009 in Russia 75 were killed by a hydro power plant turbine disintegration.

9. Avoiding weapons proliferation

Like many other technological innovations, nuclear technology was ambiguous at the outset. Its initial development was military, during World War II. Two nuclear bombs, made from uranium-235 and plutonium-239, were dropped on the Japanese cities of Hiroshima and Nagasaki respectively in August 1945 and these brought the long war to a sudden end. The immense and previously unimaginable power of the atom had been demonstrated. There was a large death toll, and survivors of the original blasts have suffered from a slightly increased incidence of cancer.

Then attention turned to civil applications. In the course of half a century, nuclear technology has enabled humankind to access a virtually unlimited source of energy at a time when constraints on the use of fossil fuels are arising.

Headquarters of the UN's International Atomic Energy Agency in Vienna (Photograph: IAEA)

The question which frames this chapter is: to what extent and in what ways does nuclear power generation contribute to or alleviate the risk of proliferation of nuclear weapons?

In the 1960s it was widely assumed that there would be 30-35 nuclear weapons states by the turn of the 20th Century. In fact there were eight – a tremendous testimony to the effectiveness of the Nuclear Non-Proliferation Treaty (NPT) and its incentives both against weapons and for a clear and unambiguous focus on civil nuclear power.

9.1 INTERNATIONAL COOPERATION TO ACHIEVE SECURITY

Nuclear weapons are now in the possession of several nations[1], and during the Cold War (1950s to 1980s) there was a massive build-up of nuclear armaments, particularly by the USA and the Soviet Union. In the last 40 years there have been strenuous international efforts to dissuade other countries from joining the five main nuclear weapons states and India[2]. These efforts have been central to the role of one particular body, the International Atomic Energy Agency (IAEA), set up in 1957 by unanimous resolution of the United Nations.

One of the main functions of the IAEA is "to establish and administer safeguards designed to ensure that special fissionable and other materials... are not used in such a way as to further any military purpose"[3]. The IAEA endeavours to detect any diversion of nuclear material from peaceful nuclear activities to be used for the

---

[1] Weapons states under the Nuclear Non-Proliferation Treaty (NPT) are the USA, UK, Russia, France, and China. Israel is described as a 'threshold state', maintaining ambiguity about its nuclear status but generally considered to have nuclear weapons capability. South Africa declared and then voluntarily dismantled a clandestine nuclear weapons program. India and Pakistan have demonstrated possession of nuclear weapons (notably through tests in May 1998) but can only properly join the NPT if they, like South Africa, voluntarily renounce and dismantle their nuclear weapons. North Korea has evidently been developing nuclear weapons. Iran is ambiguous, in pursuing uranium enrichment on a significant scale without any evident commercial justification.

[2] India is not under the NPT, though its non-proliferation bona fides are not in doubt, and for that reason by international agreement in 2008 it has achieved comparable status to NPT weapons states. See Section 9.2.

[3] IAEA Statute: Article III, paragraph 5.

Former IAEA Director Generals Hans Blix and Mohamed ElBaradei (Photograph: IAEA)

manufacture of nuclear weapons or other nuclear explosive devices. Further, it attempts to deter any such diversion by its capacity for early detection. The IAEA also advises its members on the use of nuclear materials in non-military areas such as agriculture, industry and medicine, and develops safety standards for nuclear power plants.

At the time the IAEA was being established, there was considerable concern that many countries would seek to develop or acquire nuclear weapons, just as they might upgrade their military forces with new equipment. It was in this context that the cornerstone document governing the spread of nuclear weapons, the Treaty on the Non-Proliferation of Nuclear Weapons (Non-Proliferation Treaty, or NPT), was negotiated. It entered into force in 1970. The NPT was essentially an agreement between the five nuclear weapons states and the other countries interested in nuclear technology. The deal was that assistance and cooperation would be traded for pledges, backed by international scrutiny, that no plant or material would be diverted to weapons use. Those who refused to be part of the deal would be excluded from international cooperation or trade involving nuclear technology. The NPT also represented a nuclear truce among non-weapons states, whereby they collectively resolved to turn away from the nuclear weapons option and largely disarm.

The first group of NPT signatories are non-nuclear weapons states. Each must agree not to manufacture or otherwise acquire nuclear weapons or other nuclear explosive devices. These states are obliged to conclude agreements with the IAEA for the application of safeguards on the full scope of their nuclear program (see Section 9.2).

The other NPT signatories are the five so-called nuclear weapons states. This group comprises those who had manufactured and exploded a nuclear weapon before 1967, and consists of the USA, the Soviet Union (now Russia), the United Kingdom, France and China[4]. These countries are not required to accept IAEA safeguards, although the NPT does contain certain obligations concerning disarmament which apply to them. All have, however, signed the NPT and accepted some safeguards on their peaceful nuclear activities[5]. In 2008 India was brought partly into the IAEA safeguards system, ending 34 years of trade isolation in relation to nuclear materials and technology.

The NPT was extended indefinitely in 1995. Several regional treaties complement it. Recently, other developments aimed at bolstering the non-proliferation regime have emerged. In September 1996, a Comprehensive Nuclear Test Ban Treaty was opened for signature, aimed at the elimination of nuclear weapons' testing. Negotiations are under way on a Fissile Material Cut-off Treaty, which would prohibit the further production of fissile nuclear weapons materials.

9.2 INTERNATIONAL NUCLEAR SAFEGUARDS

Over more than 40 years, the IAEA's safeguards system under the NPT has been a conspicuous international success, at least within the scope of its operation. It has involved cooperation in developing nuclear energy for electricity generation, while ensuring that civil uranium, plutonium and associated plant did not allow proliferation of nuclear weapons to occur as a result of this.

It is important to realise that international nuclear safeguards under the NPT are focused on the control of fissile materials only. They have nothing to do with engineering or organisational safety aspects of reactors, waste disposal, or transport. These are covered by other international arrangements and conventions. It is also

---

[4] France and the People's Republic of China did not ratify the NPT until 1992.
[5] The USA, UK, and France accept safeguards on all civil facilities, China and Russia on some.

Meeting of the IAEA Board (Photograph: D. Calma, IAEA)

important to understand that nuclear safeguards are a prime means of reassurance whereby non-nuclear weapons states demonstrate to others, including neighbours and rivals, that they are fulfilling their peaceful commitments. They prevent nuclear proliferation in the same way that auditing procedures build confidence in proper financial conduct and prevent embezzlement. Their specific objective is to verify whether declared (usually traded) nuclear material remains within the civil nuclear fuel cycle and is being used solely for peaceful purposes, or not. In other words, nuclear safeguards are intended to reveal whether a nation is adhering to its undertakings in relation to nuclear fuel materials. It is then up to the international community to bring pressure to bear on such a country if diversion of nuclear materials from its peaceful program is demonstrated, or other major irregularities are identified.

International nuclear safeguards are administered by the IAEA and were formally established under the NPT, which 190 states have signed. Some 70 of these have significant nuclear activities. NPT safeguards require nations to:

- Declare to the IAEA their nuclear facilities.
- Report to the IAEA what nuclear materials they hold and their location.
- Accept visits by IAEA auditors and inspectors to verify independently their material reports and physically inspect the nuclear materials concerned, to confirm physical inventories of them.

The IAEA also administers specific safeguards procedures for some countries[6] that have not joined the NPT. The

IAEA safeguards are the principal nuclear control procedures in the world today, and cover almost 900 nuclear facilities and other locations containing nuclear material in 57 non-nuclear-weapons countries. However, other safeguards systems also exist, for example amongst certain European nations (Euratom safeguards), or between individual countries (bilateral agreements) such as Australia and customer countries for its uranium, or between Japan and the USA.

These safeguards systems have been effective in preventing any diversion of materials actually covered by them. However, as nuclear power reactors, research reactors and fuel cycle components become more widespread, the safeguards task becomes more complex. At the same time, expectations have become higher. More than simply auditing declared materials is now expected under the safeguards systems, and concerns are focused on countries and activities not so far covered by them, plus Iran. Revision and upgrading of safeguards procedures is a continuing process.

For instance, Iraq showed up shortcomings in detection when it mounted an ambitious and clandestine indigenous weapons program up to 1991 which was unrelated to civil nuclear power. Discovery of this provided the impetus for a thorough reconsideration of what safeguards are expected to achieve, and how they should be implemented beyond the normal trade in civil nuclear materials and related activities. The enhanced safeguards system resulting from this will be able to provide a credible assurance that any undeclared nuclear activities would be detected in NPT countries. The focus of concern and political attention would then be squarely on countries defaulting on their international safeguards commitments (e.g. North Korea, Iraq and Iran) and also on non-NPT countries, notably Israel, Pakistan and to some extent, India.

India developed its nuclear deterrent after the NPT came into effect (rather than just before, as China), and is thus denied any full place within the NPT. It has been severely disadvantaged by the safeguards system in developing nuclear power for peaceful purposes, and special international agreements have been negotiated to address this, while acknowledging the country's good record in not passing on nuclear materials or technology to others. India's desire is to be treated the same as China.

---

[6] India, Pakistan, Israel, Cuba and Brazil. The first three have significant nuclear activities which are not subject to IAEA safeguards, although they accept them for some facilities. India has undertaken to put all civil facilities under safeguards (see also **WNA** *Nuclear Power in India* paper). Cuba and Brazil have all their nuclear activities under safeguards.

In 1997 the IAEA started to develop and implement strengthened measures, now known as 'integrated safeguards', for use by the Agency when verifying states' compliance with their commitments not to produce nuclear weapons. This would enable the IAEA to look at any nuclear-related materials and technology as possible indicators of undeclared nuclear programs, and hence of undeclared nuclear materials. These measures are detailed in an Additional Protocol to each country's agreement with IAEA, through which they accept stronger and more intrusive verification on their territory. This has now become firmly established as the standard for NPT safeguards, and most of the NPT countries with significant nuclear activities have now signed and ratified an Additional Protocol. It is hoped that all of the NPT's signatories will eventually agree to one, or at least the approximately 70 states with significant nuclear activities. In mid-2011, of the 62 non-nuclear-weapons NPT parties, 48 had an Additional Protocol in force and eight more had one signed. Those remaining without an Additional Protocol include Israel, Pakistan and North Korea – all outside the NPT anyway.

The new measures provide increased access for inspectors, both to information about current and planned nuclear programs and to more locations on the ground. Access is not restricted to declared nuclear sites, but extends almost anywhere, including high-tech industrial facilities. Inspection activity may include remote surveillance, environmental sampling and monitoring systems at key locations. States accepting the Additional Protocol will need to remove restrictive requirements on inspectors so that they can visit anywhere at short notice. In practice, this proved more of a deterrent to full take-up of the Additional Protocol than was initially envisaged.

Today many nations have the necessary trained scientists, experienced chemical technicians and the raw materials to attempt to carry out a moderate

weapons production program if they so desire. Certainly the widespread use of nuclear power for electricity generation, together with the large numbers of research reactors operating in over 50 countries, has resulted in many people being trained and experienced in aspects of nuclear technology.

The most important factor underpinning the safeguards regime is international political pressure, and how particular nations perceive their long-term security interests in relation to their immediate neighbours.

The solution to weapons proliferation is thus political more than technical, and it certainly goes beyond the question of uranium availability. International pressure not to acquire weapons is enough to deter most states from developing a weapons program. The major risk of nuclear weapons proliferation will always lie with countries which have not joined the NPT or come under its wing, and which have significant unsafeguarded nuclear activities, and those which have joined but disregard their treaty commitments. Pakistan and Israel are in the first category; North Korea and Iran are in the second[7]. While safeguards apply to some nuclear activities in non-NPT countries, other facilities and activities remain outside the scrutiny of safeguards.

Crude centrifuges found in a warehouse near Tuwaitha, Iraq, after the first Gulf war

---

[7] North Korea has subsequently withdrawn from the NPT.

9.3 FISSILE MATERIALS

Much of the concern about possible weapons proliferation arises from considering the fissile materials themselves. For instance, in relation to the plutonium contained in used fuel discharged each year from the world's commercial nuclear power reactors, it is correctly but misleadingly asserted that only a few kilograms of plutonium are required to make a bomb. Furthermore, no nation is without enough indigenous uranium to construct a few weapons (see Section 3.3).

> **There is little potential for civil nuclear power materials to be applied to weapons.**

Table 19 gives some of the important characteristics of plutonium and its use. Plutonium is a substance of varying properties depending on its source. It consists of several different isotopes, including Pu-238, Pu-239, Pu-240, and Pu-241. Not all of these are fissile – only Pu-239 and Pu-241 can undergo fission in a normal reactor. Plutonium-239 by itself is an excellent nuclear fuel, and provides about one-third of the output of a typical nuclear reactor. It has also been used extensively for nuclear weapons because it has a relatively low spontaneous fission rate and a low critical mass. Consequently plutonium-239, with only a few percent of the other isotopes present, is often called 'weapons-grade' plutonium. This was used in the Nagasaki bomb in 1945, and in many of those in world weapons stockpiles since.

On the other hand, 'reactor-grade' plutonium as routinely produced in all commercial nuclear power reactors, and which may be separated by reprocessing the used fuel from them, is not the same thing. It contains a large proportion – up to 40% – of the heavier plutonium isotopes, especially Pu-240, due to it having remained in the reactor for a relatively long time while much of the Pu-239 produced was burned up (see Figure 25). This composition is not a particular problem for re-use of the plutonium in mixed oxide (MOX) fuel for reactors (see Section 5.2), but it seriously affects the suitability of the material for nuclear weapons.

Due to spontaneous fission of Pu-240, only a very low level of it is tolerable in material for making weapons. Design and construction of nuclear explosives based on normal reactor-grade plutonium would be difficult, dangerous and unreliable, and has not so far been done[8]. However, safeguards arrangements assume that both kinds of plutonium could conceivably be used for weapons, particularly weapons designed for terror rather than military use. This is the basis of objection from some quarters to reprocessing and separation of any plutonium from used fuel.

It is worth noting that a nuclear reactor which uses mixed oxide input for one-third of its fuel is not a net producer of plutonium, and that which emerges in the used fuel is even less suitable for weapons use than what is in the fresh MOX fuel.

Figure 25. Plutonium in the reactor core

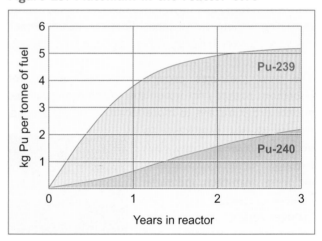

Commercial plutonium is therefore a very much less attractive material for weapons than plutonium produced in special 'production reactors'[9] designed for producing Pu-239. Such reactors need to be capable of short irradiation cycle, with frequent fuel changing. However, the development of new enrichment technology may mean that it becomes feasible to enrich commercial plutonium to weapons-grade. Hence safeguards arrangements are set up accordingly to take seriously the proliferation possibilities even of reactor-grade plutonium. (Conventional enrichment cannot readily be used to separate Pu-239 from Pu-240 because the atomic mass is so similar.)

---

[8] In 1962 a nuclear device using low-burnup plutonium from a UK Magnox reactor was detonated in the USA. The isotopic composition of this plutonium has not been officially disclosed, but it was evidently about 85% Pu-239 — what would since 1971 have been called 'fuel-grade' plutonium. The plutonium used was almost certainly from the Calder Hall or Chapelcross reactors then operating partly as military plutonium production reactors.

[9] Or heavy water-moderated research reactors, as in India.

Table 19. Plutonium

Formation

U-238 + neutron → U-239 → Np-239 → Pu-239
(Beta decay of U-239 and Np-239: 23.5 minutes and 2.35 days half-life respectively)
Pu-239 + neutron → Pu-240
Pu-240 + neutron → Pu-241

On average, one in four neutron absorptions by Pu-239 results in the formation of Pu-240 rather than in fission. Pu-241 and Pu-242 are formed by successive neutron capture in the reactor fuel. After fuel has been irradiated in the reactor for a couple of years, Pu-239 burns almost as fast as it forms, whereas Pu-240 accumulates steadily. A very small amount of Pu-238 is formed from U-235 by neutron capture.

Amount

A 1000 MWe reactor produces about 250 kg of plutonium (especially Pu-239) each year. It remains locked up in highly radioactive used fuel unless reprocessed (see Figure 16 on page 55 and Figure 17 on page 58).
The amount of Pu-240 increases with the time that fuel elements remain in the reactor (see Figure 25). Pu-240 is not fissile in a thermal reactor, but can become fissile Pu-241 by further neutron capture. (Pu-240 is fissionable in a fast neutron reactor.)

Radioactivity

Pu-239 emits alpha particles to decay to U-235 (see Appendix 2). Its half-life is 24,390 years, therefore it has a low level of radioactivity.
Pu-240 emits alpha particles as it decays to U-236 (another non-fissile isotope). Its half-life is 6600 years, therefore it has a higher level of radioactivity than Pu-239. It also emits neutrons from spontaneous fission disintegrations, as does Pu-238 (half-life 86 years).
Providing protection from this alpha radioactivity involves sealing the plutonium from physical contact e.g. in a plastic bag.

Uses

The decay heat of Pu-238 (0.56 W/g) enables its use as an energy source in the thermoelectric generators of some cardiac pacemakers, space satellites, navigation beacons etc. Plutonium power enabled the *Voyager* spacecraft to send back pictures of distant planets. Pu-240 has been used in similar applications.
The main peaceful use of Pu-239 is as nuclear reactor fuel.
Pu-241 (half-life 13 years) is the source, by beta decay, of americium-241, the vital ingredient in most household smoke detectors.

Type	Composition	Origin	Use
Reactor-grade, from high burn-up fuel	55-60% Pu-239, >19% Pu-240, typically about 30% non-fissile	Comprises about 1% of used fuel from normal operation of civil nuclear reactors used for electricity generation	As ingredient (c. 5-7%) of MOX fuel for normal reactor (can also be used as fuel in fast neutron reactor)
Weapons-grade	Pu-239 with <7% Pu-240	From military 'production' reactors specifically designed and operated for production of low burn-up Pu	Nuclear weapons (can be recycled as fuel in fast neutron reactor or as ingredient of MOX fuel)

There are two other fissile materials that could be used for weapons and both are isotopes of uranium. The most common, and the material used to make the 1945 Hiroshima bomb, is uranium-235. This material is produced by enriching natural uranium in an enrichment plant, not to 3 to 5% as required for light water reactor fuel, but to 93% U-235 or higher.

The other isotope of uranium potentially suitable for use in explosives is U-233. This does not occur naturally but is made from thorium-232 fuels in (special) reactors in much the same way as plutonium is made from U-238 in uranium-fuelled reactors (see Section 5.2). However, the use of thorium-fuelled reactors (see Sections 3.5 and 5.3) has not moved beyond the experimental stage, and U-233 is not seen as a significant proliferation problem.

Whilst the above materials can be used for explosives manufacture, they are not readily available in any practical sense, and international efforts are designed to make them even less accessible.

In recent years, the international community was challenged by an illicit nuclear weapons program in North Korea based on plutonium production in a 'research' reactor and detected by IAEA safeguards inspections. The United Nations imposed a nuclear 'freeze' on the country's reactors and facilities under the 1994 Agreed Framework between North Korea and the United States. This led to North Korea bowing to international pressure so that the IAEA could reassure the UN that all nuclear materials were safeguarded and that North Korea was moving towards full compliance with its IAEA safeguards agreement. The trade-off for North Korea was that an international consortium led by the USA, South Korea and Japan started building two large modern nuclear reactors for the country to provide electricity untainted by military possibilities. However, in 2002 North Korea admitted to a clandestine uranium enrichment program, which put the country doubly in default of its international treaty obligations. It then left the NPT, placing itself outside the IAEA safeguards regime, and little has been resolved since.

Even greater concern was generated by suspicions that Iraq had developed or was developing nuclear weapons; these fears were heightened during the 1991 Persian Gulf War. After the ceasefire in 1991, the United Nations was able to confirm that Iraq, though a signatory to the NPT, had been pursuing a clandestine weapons program quite separately from materials and facilities covered by IAEA inspections. The major part of its illicit endeavour was based on indigenous uranium and its enrichment. As noted

in Section 9.2, this situation led to the enhancement of the safeguards regime, and the IAEA's Additional Protocol.

Iran attracted world attention in 2002 when previously undeclared nuclear facilities became the subject of IAEA inquiry. On investigation, the IAEA found inconsistencies in Iran's declarations to the Agency and raised questions as to whether Iran was in violation of its safeguards agreement, as a signatory of the NPT. An IAEA report released to its member states in November 2003 showed that, in a series of contraventions of its safeguards agreement over 22 years, Iran had systematically concealed its development of key techniques – notably uranium enrichment – which are capable of use for non-peaceful purposes. Iran has not convinced the IAEA or the UN Security Council that its uranium enrichment is legitimate, and the situation remains unresolved as of 2012.

9.4 RECYCLING MILITARY URANIUM AND PLUTONIUM FOR ELECTRICITY

International efforts aimed at nuclear disarmament have, ironically, led to some serious safety and security problems. Dismantling of nuclear warheads under US-Russia disarmament agreements (START I and START II) has resulted in an accumulation of weapons-grade material (plutonium and high-enriched uranium). Concerns have arisen, particularly following the break-up of the Soviet Union, about the possibility that these fissile materials could be subject to theft, smuggling, or illicit trafficking, and could make their way into the hands of rogue states or terrorists. Inadequate control of nuclear materials inside Russia, the sheer size of Russian nuclear programs, and substandard security at nuclear installations are a few of the factors that have raised concerns about nuclear materials falling into the wrong hands. The joint efforts of many nations through the 1990s and since has considerably improved the physical security and accountability of such materials.

The prospect of using weapons-grade plutonium (more than about 93% Pu-239) in mixed oxide (MOX) fuel for civil reactors is receiving increased attention. It is quite feasible to make MOX fuel using a mixture of military and reactor-grade plutonium with depleted uranium, and some has been made with military plutonium only. Burning it in MOX fuel is currently the only practical means of disposal which permanently removes military plutonium from circulation and effectively destroys it. There are current initiatives to recycle plutonium in this manner.

So, after three decades of concern regarding the possibility of uranium intended for commercial nuclear power finding its way into weapons, we have now seen a large amount of military uranium being directed into the civil nuclear fuel cycle, for use in commercial nuclear power generation. The first such material from Soviet military warheads arrived in the USA in 1995, and it now provides 10% of all US electricity. A start has also been made on recycling US weapons-grade uranium for electricity. Military high-enriched uranium is diluted about 25:1 with depleted uranium left over from enrichment plants, or similar material (see also Section 3.4).

9.5 AUSTRALIAN AND CANADIAN NUCLEAR SAFEGUARDS POLICIES

Canada and Australia produce almost half of the world's mined uranium and they export nearly all of that, so therefore provide a case study in how non-proliferation is approached. Both countries are strong proponents of a robust international non-proliferation regime to enhance national and international security. Both are rigorous in seeking assurances that nuclear exports will only be used for legitimate and peaceful nuclear energy purposes.

Australia's main interest in international nuclear safeguards is in relation to the use of its uranium in overseas nuclear power programs. Canada's interest is broader, covering the whole domestic fuel cycle, plus the export of both uranium and reactor technology. In both countries, exports of uranium are controlled by the federal government.

Following World War II, Canada pledged that it would not develop nuclear weapons, even though it had, at the time, the capability to do so. Both Canada and Australia participated in the drafting of the Statute of the IAEA, have been continuously represented on the IAEA's Board of Governors, and remain active in many of the various technical committees and advisory groups of the IAEA.

After a major public inquiry, the Australian Government decided on the basic principles of a national safeguards policy, and these were announced during 1977. Australia was involved in the International Nuclear Fuel Cycle Evaluation program in the 1970s and continues to use its status as a uranium supplier to press for high safeguards standards to be applied.

Table 20 sets out in summary the main elements of both countries' policies.

Table 20. Australian and Canadian nuclear safeguards policies

1. **Selected countries**
    Non-weapons states must be party to the NPT and must accept full-scope IAEA safeguards applying to all their nuclear-related activities. Australia requires them to have ratified the Additional Protocol to their safeguards agreement with the IAEA.
    Weapons states to give assurance of peaceful use; IAEA safeguards to cover the material.

2. **Bilateral agreements are required**
    IAEA to monitor compliance with IAEA safeguards requirements.
    Fallback safeguards (if NPT ceases to apply or IAEA cannot perform its safeguards functions).
    Prior consent required to transfer material or technology to another country.
    Prior consent required to enrich above 20% U-235.
    Prior consent required to reprocess.
    Control over storage of any separated plutonium.
    Adequate physical security.

3. **Materials exported or re-exported to be in a form attracting full IAEA safeguards**

4. **Commercial contracts to be subject to conditions of bilateral agreements**

5. **Australia and Canada will participate in international efforts to strengthen safeguards**

6. **Australia and Canada recognise the need for constant review of standards and procedures**

The Australian and Canadian policies outlined in Table 20 are based on the requirements of the NPT and the IAEA safeguards invoked under it. Superimposed on these are conditions which are required by bilateral agreement with customer countries[10] and implemented by the Australian Safeguards and Non-proliferation Office (ASNO) or the Canadian Nuclear Safety Commission (CNSC) respectively. Both countries' legally-binding bilateral safeguards measures are directed towards preventing any unauthorised or clandestine use of exported uranium or any materials derived from it: 'Australian-obligated nuclear materials'[11] or the Canadian equivalent. The Canadian agreements cover nuclear material, heavy water, nuclear equipment and technology. The bilateral safeguards are designed to deter possible diversion of fissile material or misuse of equipment and technology, and to do so more rigorously than standard IAEA safeguards on their own.

Canada's CNSC is responsible for regulating domestic nuclear facilities and is charged with administering the agreement between Canada and the IAEA for the application of safeguards in Canada. The Commission assists the IAEA by allowing access to Canadian nuclear facilities and arranging for the installation of safeguards equipment at the sites. It also reports regularly to the IAEA on nuclear materials held in Canada. The CNSC also manages a program for research and development in support of IAEA safeguards, the Canadian Safeguards Support Program.

Australia's ASNO performs a similar role, apart from the regulation (which is undertaken by another federal authority). It administers the safeguards agreement with the IAEA, arranges IAEA access to Australian facilities, and reports to the IAEA on nuclear materials in Australia.

See also: WNA information paper on *Safeguards to Prevent Nuclear Proliferation*.

---

[10] Australia has 21 bilateral safeguards agreements covering 46 countries (the Euratom agreement covering all 27 countries in the EU, giving some overlap); Canada has 20 agreements in force, including with Euratom.

[11] AONM comprises all Australian-origin material in the front end of the fuel cycle, including depleted uranium, and also the uranium and plutonium in used fuel at the back end.

10. History of nuclear energy

The history of nuclear energy starts with science in continental Europe, then blossoms in the UK and USA with the latter's technological might, languishes for a few decades, then has a new growth spurt particularly in east Asia.

Uranium, the key element in the story, was discovered in 1789 by Martin Klaproth, a German chemist, and named after the planet Uranus. The knowledge of atomic radiation, atomic change and nuclear fission was developed much later, from 1895 to 1945, and much of it in the last six of those years.

10.1 EXPLORING THE NATURE OF THE ATOM

Ionising radiation was studied and described by Wilhelm Röntgen in 1895, after he passed an electric current through an evacuated glass tube and produced X-rays. Then in 1896 Henri Becquerel found that pitchblende (an ore containing radium and uranium) caused a photographic plate to darken. He went on to demonstrate that this was due to beta radiation (electrons) and alpha particles (helium nuclei) being emitted from the mineral itself. Paul Ulrich Villard found a third type of radiation from pitchblende: gamma rays, which were much the same as X-rays. Then in 1896, Pierre and Marie Curie gave the name 'radioactivity' to this phenomenon, and in 1898 they isolated polonium and radium from the pitchblende. Radium was later used in medical treatment. In 1898 Samuel Prescott showed that radiation destroyed bacteria in food.

In 1902 Ernest Rutherford showed that radioactivity as a spontaneous event involving the emission of an alpha or beta particle from the nucleus of an atom created a different element. In 1919 he fired alpha particles from a radium source into nitrogen and found that nuclear rearrangement was occurring, with formation of oxygen. Through to the 1940s, Niels Bohr advanced our understanding of the way electrons were arranged around the atom's nucleus.

By 1911 Frederick Soddy had discovered that naturally-radioactive elements had a number of different isotopes (radionuclides), with the same chemistry. Also in 1911, George de Hevesy showed that such radionuclides were invaluable as tracers, because minute amounts could readily be detected with simple instruments.

In 1932 James Chadwick discovered the neutron. In the same year, Cockcroft and Walton produced nuclear transformations by bombarding atoms with accelerated protons. Then in 1934 Irene Curie and Frederic Joliot found that some such transformations created artificial radionuclides. The next year Enrico Fermi found that a much greater variety of artificial radionuclides could be formed when neutrons were used instead of protons. Fermi continued his experiments, mostly producing heavier elements from his targets, but also, with uranium, some much lighter ones. At the end of 1938 Otto Hahn and Fritz Strassmann in Berlin showed that the new lighter elements were barium and others which were about half the mass of uranium, thereby demonstrating that atomic fission had occurred. Lise Meitner and her nephew Otto Frisch, working under Niels Bohr, then explained this

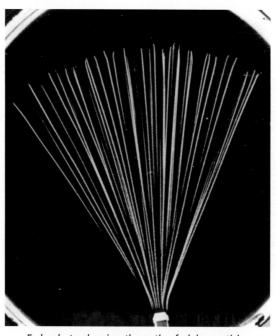
Early photo showing the path of alpha particles in a cloud chamber

by suggesting that the neutron was captured by the nucleus, causing severe vibration leading to the nucleus splitting into two not quite equal parts. They calculated the energy release from this fission as about 200 million electron volts. Frisch then confirmed this figure experimentally in January 1939. This was the first experimental confirmation of Albert Einstein's paper putting forward the equivalence between mass and energy, which had been published in 1905.

10.2 HARNESSING NUCLEAR FISSION

The developments in 1939 sparked activity in many laboratories. Hahn and Strassmann showed that fission not only released a lot of energy but that it also released additional neutrons which could cause fission in other uranium nuclei and possibly a self-sustaining chain reaction leading to an enormous release of energy. This suggestion was soon confirmed experimentally by Joliot and his co-workers in Paris, and also by Leo Szilard working with Fermi in New York.

Bohr soon proposed that fission was much more likely to occur in the uranium-235 isotope than in U-238 and that fission would occur more effectively with slow-moving neutrons than with fast neutrons. The latter point was confirmed by Szilard and Fermi, who proposed using a 'moderator' to slow down the emitted neutrons. Bohr and Wheeler extended these ideas into what became the classical analysis of the fission process, and their paper was published only two days before World War II broke out in 1939.

Another important factor was that U-235 was known to comprise only 0.7% of natural uranium, with the other 99.3% being U-238. The fact that different isotopes (of the same element) have the same chemical properties makes their separation difficult and requires the use of their very slightly different physical properties. In this case increasing the proportion of the U-235 isotope gave rise to several different processes known as 'enrichment'.

The remaining piece of the fission concept was provided in 1939 by Francis Perrin who introduced the concept of the critical mass of uranium required to produce a self sustaining release of energy. His theories were extended by Rudolf Peierls at Birmingham University, and the resulting calculations were of considerable importance in the development of the atomic bomb. Perrin's group in Paris continued their studies and demonstrated that a chain reaction could be sustained in a uranium-water mixture (the water being used to slow down the neutrons) provided external neutrons were injected into the system. They also demonstrated the idea of introducing neutron absorbing material to limit the multiplication of neutrons and thus control the nuclear reaction (which is the basis for the operation of a nuclear power station).

Peierls had been a student of Werner Heisenberg, who from April 1939 presided over the German nuclear energy project (*Uranprojekt*, also known as the *Uranverein*, or Uranium Club) under the German Ordnance Office. Initially this was directed towards military applications, but by 1942 the military objective was abandoned as impractical. However, the existence of the German *Uranverein* project provided the main incentive for wartime development of the atomic bomb by Britain and the USA.

7th Solvay conference in Brussels in 1933. Persons: Kramers, Hendrik Anton; Mott, Neville Francis; Gamow, George; Blackett, Patrick Maynard Stuart; Cosyns, M.; Piccard, Aug.; Stahel, E.; Dirac, Paul Adrian Maurice; Errera, J.; Ellis, Charles Drummond; Lawrence, Ernest Orlando; Henriot, E.; Joliot-Curie, Frederic; Heisenberg, Werner Karl; Walton, E.T.S.; Debye, Peter; Cabrera, B.; Bothe, Walther William; Bauer, H.E.G.; Verschaffelt, J.E.; Cockcroft, John Douglas; Rosenfeld, Lèon; Perrin, F.; Fermi, Enrico; Rosenblum, M. Salomon; Pauli, Wolfgang; Herzen, E.; Peierls, Rudolf Ernst; Schrödinger, Erwin; Joliot-Curie, Irene; Bohr, N.; Joffe, Abram Feodorovich; Curie, Marie; Richardson, Owen Williams; Rutherford, Ernest; Broglie, Maurice de; Meitner, Lise; Chadwick, James; Langevin, Paul; Donder, Th. de; Broglie, Louis Victord. (Photo courtesy of Inst. Int. de Physique Solvay)

10.3 NUCLEAR PHYSICS IN RUSSIA

Russian nuclear physics predates the Bolshevik Revolution by more than a decade. Work on radioactive minerals found in central Asia began in 1900 and the St Petersburg Academy of Sciences began a large-scale investigation in 1909. The 1917 Revolution gave a boost to scientific research and over ten physics institutes were established in major Russian towns, particularly St Petersburg, in the years which followed. In the 1920s and early 1930s many prominent Russian physicists worked abroad, encouraged by the new regime initially as the best way to raise the level of expertise quickly. These included Kirill Sinelnikov, Pyotr Kapitsa and Vladimir Vernadsky.

By the early 1930s there were several research centres specialising in nuclear physics. Sinelnikov returned from Cambridge in 1931 to organise a department at the Kharkov Institute of Physics and Technology (KIPT), which had been set up in 1928 as the Ukrainian Institute of Physics and Technology. Academician Abram Ioffe formed another group (including the young Igor Kurchatov) at the Leningrad Physical Technical Institute (now the Ioffe Institute), which in 1933 became the Department of Nuclear Physics under Kurchatov with four separate laboratories.

By the end of the decade, there were cyclotrons installed at the Radium Institute and the Ioffe Institute (the biggest in Europe) in Leningrad. But by this time many scientists were beginning to fall victim to Stalin's purges – half the staff of KIPT, for instance, was arrested in 1939. Nevertheless, 1940 saw great advances being made in the understanding of nuclear fission including the possibility of a chain reaction. At the urging of Kurchatov and his colleagues, the Academy of Sciences set up a 'Committee for the Problem of Uranium' in June 1940 chaired by Vitaly Khlopin, and a fund was established to investigate the central Asian uranium deposits. With Germany's invasion of Russia in 1941, much of the research switched to potential military applications.

10.4 CONCEIVING THE ATOMIC BOMB

In Britain, the refugee physicists Rudolf Peierls and Otto Frisch (who stayed in England with Peierls after the outbreak of war) gave a major impetus to the concept of an atomic bomb in a three-page document known as the *Frisch-Peierls Memorandum*. In this they predicted that

Three leading Soviet nuclear physicists (from left to right) Abram Ioffe, Abram Alixanov, Igor Kurchatov

an amount of about 5 kg of pure U-235 could make a very powerful atomic bomb equivalent to several thousand tonnes of dynamite. They also suggested how such a bomb could be detonated, how the U-235 could be produced, and what the radiation effects might be in addition to the explosive effects. They proposed thermal diffusion as a suitable method for separating the U-235 from the natural uranium. This memorandum stimulated a considerable response in Britain at a time when there was little interest from the USA.

A group of eminent scientists known as the MAUD[1] Committee was set up in Britain and supervised research at the Universities of Birmingham, Bristol, Cambridge,

---

[1] Military Application of Uranium Detonation

Liverpool and Oxford. The chemical problems of producing gaseous compounds of uranium and pure uranium metal were studied at Birmingham University and by Imperial Chemical Industries (ICI). Dr Philip Baxter[2] at ICI made the first small batch of gaseous uranium hexafluoride for Professor James Chadwick in 1940. ICI received a formal contract later in 1940 to make 3 kg of this vital material for the future work. Most of the other research was funded by the universities themselves.

Two important developments came from the work at Cambridge. The first was experimental proof that a chain reaction could be sustained with slow neutrons in a mixture of uranium oxide and heavy water *i.e.* the output of neutrons was greater than the input. The second arose from work by Egon Bretscher and Norman Feather and was based on earlier research by Hans von Halban and Lew Kowarski soon after they arrived in Britain from Paris. In 1937 Halban had been invited to join the team of Frédéric Joliot-Curie at the Collège de France in Paris, with Francis Perrin and Lew Kowarski. In 1939 the

Rudolph Peierls – co-author of the Frisch-Peierls Memorandum

group measured the mean number of neutrons emitted during nuclear fission, and established the possibility of nuclear chain reactions and nuclear energy production. In August the group showed that the rate of fission in uranium oxide was increased by immersion in ordinary (light) water.

When U-235 and U-238 absorb slow neutrons, the probability of fission in U-235 is much greater than in U-238. The U-238 is more likely to form a new isotope U-239, and this isotope rapidly emits an electron to become a new element with a mass of 239 and an atomic number of 93. This element also emits an electron and becomes a new element of mass 239 and atomic number 94, which has a much greater half life. Bretscher and Feather argued on theoretical grounds that element 94

would be readily fissionable by slow and fast neutrons, and had the added advantage that it was chemically different to uranium and therefore could easily be separated from it.

This new development was also confirmed in independent work by Edwin McMillan and Philip Abelson at the Berkeley Radiation Laboratory at the University of California, Berkeley in 1940. Dr Nicholas Kemmer of the Cambridge team proposed the names neptunium and plutonium for the new elements 93 and 94 by analogy with the outer planets Neptune and Pluto beyond Uranus (uranium, element 92). The Americans fortuitously suggested the same names, and the identification of plutonium in 1941 is generally credited to Glenn Seaborg at Berkeley.

10.5 DEVELOPING THE CONCEPTS: BOMB AND BOILER

By the end of 1940 remarkable progress had been made by the several groups of scientists coordinated by the MAUD Committee, with the expenditure of a relatively small amount of money. All of this work was kept secret, whereas in the USA several publications continued to appear in 1940 and there was also little sense of urgency.

By March 1941 one of the most uncertain pieces of information was confirmed – the fission cross-section (a measure of the probability of nuclear fission occurring) of U-235. Peierls and Frisch had initially predicted in 1940 that almost every collision of a neutron with a U-235 atom would result in fission, and that both slow and fast neutrons would be equally effective. It was later discerned that slow neutrons were very much more effective, which was of enormous significance for nuclear reactors but fairly academic in the bomb context. Peierls then stated that there was now no doubt that the whole scheme for a bomb was feasible provided highly enriched U-235 could be obtained. The predicted critical size for a sphere of U-235 metal was about 8 kg, which might be reduced by use of an appropriate material for reflecting neutrons. However, direct measurements on U-235 were still necessary and the British pushed for urgent production of a few micrograms.

[2] Dr Baxter later was sent to the Oak Ridge Laboratory in the USA to assist in the operation of the large enrichment plant secretly constructed to make the material for the first atomic bombs. He later became a key figure in the Australian Atomic Energy Commission.

The final outcome of the MAUD Committee was two summary reports in July 1941. One was on *Use of Uranium for a Bomb* and the other was on *Use of Uranium as a Source of Power*. The first report concluded that a bomb was feasible and that one containing some 12 kg of active material would be equivalent to 1800 tons of TNT; it would moreover release large quantities of radioactive

James Chadwick: Discoverer of the neutron and member of the MAUD Committee

substances, which would make places near the explosion site dangerous to humans for a long period. It estimated that a plant to produce 1 kg of U-235 per day would cost £5 million and would require a large skilled labour force that was also needed for other parts of the war effort. Suggesting that the Germans could also be working on the bomb, it recommended that the work should be continued with high priority in cooperation with the Americans, even though they seemed to be concentrating on the future use of uranium for power and naval propulsion.

The second MAUD report concluded that the controlled fission of uranium could be used to provide energy in the form of heat for use in machines in industrial applications, as well as providing large quantities of radioisotopes which could be used as substitutes for radium. It referred to the use of heavy water and possibly graphite as moderators for the fast neutrons, and that even ordinary water could be used if the uranium was enriched in the U-235 isotope. It concluded that the 'uranium boiler' had considerable promise for future peaceful uses but that it was not worth considering during the present war. The Committee recommended that Halban and Kowarski should move to the USA, where there were plans to make heavy water on a large scale. The possibility that the new element plutonium might be more suitable than U-235 was mentioned, and that therefore the work in this area by Bretscher and Feather (see Section 10.4) should be continued in Britain.

The two reports led to a complete reorganisation of work on the bomb and the 'boiler'. It was claimed that the work of the committee had put the British in the lead and it was claimed that "in its fifteen months' existence it had proved itself one of the most effective scientific committees that ever existed." The basic decision that the bomb project would be pursued urgently was taken by the Prime Minister, Winston Churchill, with the agreement of the Chiefs of Staff.

The reports also led to high-level reviews in the USA, particularly by a Committee of the National Academy of Sciences, initially concentrating on the nuclear power aspect. Little emphasis was given to the bomb concept until 7 December 1941, when the Japanese attacked Pearl Harbour and the Americans entered the War directly. The huge resources of the USA were then applied without reservation to developing atomic bombs.

10.6 THE MANHATTAN PROJECT

The Americans increased their effort rapidly and soon outstripped the British. Research continued in each country with some exchange of information. Several of the key British scientists visited the USA early in 1942 and were given full access to all of the information available. The Americans were pursuing three enrichment processes in parallel: Professor Ernest Lawrence was studying electromagnetic separation at Berkeley (University of California); Eger V. Murphree of Standard Oil was studying the centrifuge method developed by Professor Jesse Beams; and Professor Harold Urey was coordinating the gaseous diffusion work at Columbia University. Responsibility for building a reactor to produce fissile plutonium was given to Arthur Compton at the University of Chicago. The British were only examining gaseous diffusion for enriched uranium.

In June 1942 the US Army took over process development, engineering design, procurement of materials and site selection for pilot plants for four methods of making fissionable material (because none of the four had been shown to be clearly superior at that point) as well as the production of heavy water. With this change, information flow to Britain dried up. This was a major setback to the British and the Canadians who had been collaborating on heavy water

Enrico Fermi directed the team which produced the first controlled nuclear chain reaction in 1942

production and on several aspects of the research programme. Thereafter, Churchill sought information on the cost of building a diffusion plant, a heavy water plant and an atomic reactor in Britain.

After many months of negotiations an agreement was finally signed by Churchill and President Roosevelt in Quebec in August 1943, according to which the British handed over all of their reports to the Americans and in return received copies of General Groves' progress reports to the President. The latter showed that the entire US programme would cost over $1000 million – all for the bomb, as no work was being done on other applications of nuclear energy.

Construction of enrichment plants for electromagnetic separation (in calutrons) and gaseous diffusion was well

under way. An experimental graphite pile constructed by Enrico Fermi had operated at the University of Chicago in December 1942 – the first controlled nuclear chain reaction.

A full-scale production reactor for plutonium was being constructed at Argonne, with further ones at Oak Ridge and then Hanford, plus a reprocessing plant to extract the plutonium. Four plants for heavy water production were being built, one in Canada and three in the USA. A team under Robert Oppenheimer at Los Alamos in New Mexico was working on the design and construction of both U-235 and Pu-239 bombs. The outcome of the huge effort, with assistance from the British teams, was that sufficient Pu-239 and highly enriched U-235 (from calutrons and diffusion at Oak Ridge) was produced by mid-1945. The uranium mostly originated from the Belgian Congo.

The first atomic device tested successfully at Alamagordo in New Mexico on 16 July 1945. It used plutonium made in a nuclear pile. The teams did not consider that it was necessary to test a simpler U-235 device. The first atomic bomb, which contained U-235, was dropped on Hiroshima on 6 August 1945. The second bomb, containing Pu-239, was dropped on Nagasaki on 9 August. That same day, the USSR declared war on Japan. On 10 August 1945 the Japanese Government surrendered.

10.7 THE SOVIET BOMB

Initially Stalin was not enthusiastic about diverting resources to develop an atomic bomb, until intelligence reports suggested that such research was under way in Germany, Britain and the USA. Consultations with Academicians Ioffe, Kapitsa, Khlopin and Vernadsky convinced him that a bomb could be developed relatively quickly and he initiated a modest research programme in 1942. Igor Kurchatov, then relatively young and unknown, was chosen to head it and in 1943 he became Director of Laboratory No. 2 recently established on the outskirts of Moscow. This was later renamed Laboratory for Measuring Instruments (LIPAN), and then the Kurchatov Institute of Atomic Energy. Overall responsibility for the bomb programme rested with Security Chief Lavrenti Beria and its administration was undertaken by the First Main Directorate (later called the Ministry of Medium Machine Building).

Research had three main aims: to achieve a controlled chain reaction; to investigate methods of isotope separation; and to look at designs for both enriched uranium and plutonium bombs. Attempts were made to initiate a chain reaction using two different types of atomic pile: one with graphite as a moderator and the other with heavy water. Three possible methods of isotope separation were studied: counter-current thermal diffusion, gaseous diffusion and electromagnetic separation.

After the defeat of Nazi Germany in May 1945, German scientists were 'recruited' to the bomb programme to work in particular on isotope separation to produce enriched uranium. This included research into gas centrifuge technology in addition to the three other enrichment technologies.

The test of the first US atomic bomb in July 1945 had little impact on the Soviet effort, but by this time, Kurchatov was making good progress towards both a uranium and a plutonium bomb. He had begun to design an industrial scale reactor for the production of plutonium, while those scientists working on uranium isotope separation were making advances with the gaseous diffusion method.

It was the bombing of Hiroshima and Nagasaki the following month which gave the programme a high profile and construction began in November 1945 of a new city in the Urals which would house the first plutonium production reactors – Chelyabinsk-40 (later known as Chelyabinsk-65 or the Mayak Production Association, and now known as Ozersk). This was the first of ten closed and secret nuclear cities to be built in the Soviet Union. The first of five reactors at Chelyabinsk-65 came on line in 1948. This town also housed a processing plant for extracting plutonium from irradiated uranium.

As for uranium enrichment technology, it was decided in late 1945 to begin construction of the first gaseous diffusion plant at Verkh-Neyvinsk (later the closed city of Sverdlovsk-44 and now known as Novouralsk), some 50 kilometres from Yekaterinburg (formerly Sverdlovsk) in the Urals. Special design bureaux were set up at the Leningrad Kirov Metallurgical and Machine-Building Plant and at the Gorky (Nizhny Novgorod) Machine-Building Plant. Support was provided by a group of German scientists working at the Sukhumi Physical Technical Institute.

In April 1946 design work on the bomb was shifted to the Design Bureau No.11 (KB-11) – at a new centre at Sarova some 400 kilometres from Moscow (subsequently the closed city of Arzamas-16, now the Russian Federal Nuclear Center – The All-Russian Research Institute of Experimental Physics, RFNC – VNIIEF). More specialists were brought into the programme including metallurgist Yefim Slavsky who was given the immediate task of producing the very pure graphite Kurchatov needed for his plutonium production pile constructed at Laboratory No. 2 known as F-1 ('Physics-1'). The pile was operated for the first time in December 1946. Support was also given by Laboratory No. 3 in Moscow – now the Institute of Theoretical and Experimental Physics (ITEP) – which had been working on nuclear reactors.

Work at Arzamas-16 was influenced by foreign intelligence gathering and the first device was based closely on the Nagasaki bomb (a plutonium device). In August 1947 a test site was established near Semipalatinsk in Kazakhstan and was ready for the detonation two years later of the first bomb, RDS-1. Even before this was tested in August 1949, another group of scientists led by Igor Tamm and including Andrei Sakharov had begun work on a hydrogen bomb.

Replica of the first Soviet atomic bomb in the Russian Federal Nuclear Centre Museum

10.8 REVIVAL OF THE 'NUCLEAR BOILER'

By the end of World War II, the project predicted and described in detail only five and a half years before in the Frisch-Peierls Memorandum had been brought to partial fruition, and attention could now turn to the peaceful and directly beneficial application of nuclear

One of the world's first nuclear power reactors, Calder Hall in the UK, operated for nearly 50 years

energy. Post-war, weapons development continued on both sides of the 'Iron Curtain', but a new focus was on harnessing atomic power for making steam and electricity.

In the course of developing nuclear weapons, the Soviet Union and the West had acquired a range of new technologies, and scientists realised that the tremendous heat produced in the process could be tapped either for direct use or for generating electricity. It was also clear that this new form of energy would allow development of compact long-lasting power sources which could have various applications, not least for shipping, and especially in submarines.

The first nuclear reactor to produce electricity (albeit a trivial amount) was the small Experimental Breeder Reactor (EBR-1) in Idaho, in the USA, which started up in December 1951.

In 1953 President Eisenhower proposed his 'Atoms for Peace' programme, which reoriented significant research

effort towards electricity generation and set the course for civil nuclear energy development in the USA.

In the Soviet Union, work was under way at various centres to refine existing reactor designs and develop new ones. The existing graphite-moderated channel-type plutonium production reactor was modified for heat and electricity generation and in 1954 the world's first nuclear-powered electricity generator began operation in the then closed city of Obninsk at a newly-established branch of the Moscow Engineering and Physics Institute (now the Obninsk Institute for Nuclear Power, part of the National Research Nuclear University, MEPhI). The AM-1 (*Atom Mirny* – peaceful atom) reactor was water-cooled and graphite-moderated, with a design capacity of 30 MWt or 5 MWe. It was similar in principle to the plutonium production reactors in the closed military cities and served as a prototype for other graphite channel reactor designs including the RBMK (*reaktor bolshoi moshchnosty kanalny* – high power channel reactor) reactors. AM-1 produced electricity until 1959 and was used until 2000 as a research facility and for the production of isotopes.

Also in the 1950s, Obninsk was developing fast breeder reactors (FBRs). In 1955 the BR-1 (*bystry reaktor* – fast reactor) fast neutron reactor began operating. It produced no power but led directly to the BR-5, which started up in 1959 with a capacity of 5 MWt. BR-5 was used for basic research in designing sodium-cooled FBRs. It was upgraded and modernised in 1973 and then underwent major reconstruction in 1983 to become the BR-10 with a capacity of 8 MWt which was then used to investigate fuel endurance, study materials, and also to produce isotopes.

Admiral Hyman Rickover presided over the main US effort, which developed the pressurized water reactor (PWR) for naval (particularly submarine) use. The PWR used enriched uranium oxide fuel and was moderated and cooled by ordinary (light) water. The Mark I prototype naval reactor started up in March 1953 in

Idaho, and the first nuclear-powered submarine, *USS Nautilus*, was launched in 1954 (see Section 7.5). In 1959 both the USA and the USSR launched their first nuclear-powered surface vessels.

The Mark 1 reactor led to the US Atomic Energy Commission building the 60 MWe Shippingport demonstration PWR reactor in Pennsylvania, which started up in 1957 and operated until 1982. It was the first full-scale civil nuclear power reactor.

Since the USA had a virtual monopoly on uranium enrichment in the West, British nuclear power development took a different tack and resulted in a series of reactors fuelled by natural uranium metal, moderated by graphite, and gas-cooled. The first of these 50 MWe Magnox types, Calder Hall 1, started up in 1956 and ran until 2003. However, after 1963 (and 26 units) no more were commenced. Britain next embraced the advanced gas-cooled reactor (using enriched oxide fuel) before conceding the pragmatic virtues of the PWR design.

10.9 NUCLEAR ENERGY GOES COMMERCIAL

In the USA, Westinghouse designed the first fully commercial PWR – Yankee Rowe (250 MWe), which started up in 1960 and operated to 1992. Meanwhile the boiling water reactor (BWR) was developed by the Argonne National Laboratory, and the first one, Dresden 1 of 250 MWe, designed by General Electric, was started up earlier in 1960. A prototype BWR, Vallecitos, ran from 1957 to 1963. By the end of the 1960s, orders were being placed for PWR and BWR reactor units of more than 1000 MWe.

Canadian reactor development headed down a quite different track, using natural uranium fuel and heavy water as a moderator and coolant. The first unit started up in 1962. This 'CANDU' (Canada deuterium uranium) design has been exported, and continues to be refined.

France started out with a gas-graphite design similar to Magnox and the first reactor started up in 1956. Commercial models operated from 1959. It then settled on three successive generations of standardised PWRs, which was a very cost-effective strategy.

In 1964 the first two Soviet nuclear power plants were commissioned. A 100 MWe boiling water graphite

The first commercial US nuclear power plant: Shippingport (1957-82)

channel reactor began operating in Beloyarsk (Urals). In Novovoronezh (Volga region) a new design – a small (210 MWe) pressurised water reactor known as a VVER (*veda-vodyanoi energetichesky reaktor* – water moderated, water cooled power reactor) was built.

The first large (1000 MWe) RBMK started up at Sosnovy Bor near Leningrad in 1973. In the Arctic northwest a slightly bigger VVER with a rated capacity of 440 MWe began operating and this became a standard design, subsequently enlarged to 1000 MWe.

In Kazakhstan the world's first commercial prototype fast neutron reactor (the BN-350) started up in 1972, with 120 MW of electric power and also producing heat to desalinate Caspian sea water. In the USA, UK, France and Russia a number of experimental fast neutron reactors produced electricity from 1959, the last of these closing in 2009. This left Russia's BN-600 at Beloyarsk as the only commercial fast reactor.

Around the world, with few exceptions, other countries have chosen light-water designs for their nuclear power programmes, so that today 60% of world capacity is PWR and 21% BWR.

10.10 THE NUCLEAR POWER RENAISSANCE

From the late 1970s to about 2002 the nuclear power industry suffered some decline and stagnation. The 1979 accident at Three Mile Island in the USA eroded confidence in the technology (despite not harming anyone), and the 1986 Chernobyl disaster gave rise to great public concern, although the particular RBMK design made it technically irrelevant to most of the

The first EPR under construction at Olkiluoto in Finland (Photo: Hannu Huovila/TVO)

world's power reactors. Many reactor orders from the 1970s were cancelled. The few new reactors that were ordered, coming on line from the mid-1980s, little more than matched retirements. Against this, capacity increased by nearly one-third and output increased 60% due to capacity uprates plus improved load factors; therefore the share of nuclear in world electricity from the mid-1980s was fairly constant at 16-17%.

The industry's stagnation together with an increase in secondary supplies of uranium led to a drop in the uranium price. Oil companies which had entered the uranium field bailed out, and there was a consolidation of uranium producers.

By the late 1990s, an expansion in nuclear power in Asian countries such as Japan and South Korea ran counter to the trend in the rest of the world, and China started its nuclear power expansion. The first third-generation reactor – Kashiwazaki-Kariwa 6, a 1350 MWe advanced BWR – was commissioned in Japan in 1996-97.

In the new century, several factors have combined to revive the prospects for nuclear power. First is the realisation of the scale of projected increased electricity demand worldwide, but particularly in rapidly-developing countries. Secondly is awareness of the importance of energy security – the prime importance of each country having assured access to affordable energy. Thirdly is the need to limit carbon emissions due to concern about climate change. And fourthly, where fossil fuel prices have increased strongly or threaten to do so, the economic competitiveness of nuclear power is enhanced.

These factors coincide with the availability of a new generation of nuclear power reactors, and in 2004 the first of the later third-generation units was ordered for Finland – a 1600 MWe European Pressurized Water Reactor (EPR). It is now under construction, albeit with major cost and schedule overruns. A similar unit is under construction in France as the first of a possible full fleet replacement there. In the USA the 2005 Energy Policy Act provided encouragement for establishing new-generation power reactors and by early 2012 the first four large new Westinghouse AP1000 power reactors were being built, with seven more firmly planned to follow.

But plans in Europe and North America are overshadowed by those in China, India, and South Korea. China alone plans a huge increase in nuclear power capacity by 2020, and has more than 100 further large units proposed and backed by credible political determination and popular support. A large portion of these are the latest Western design, expedited by modular construction. Despite the political effects of the Fukushima accident in March 2011, nuclear power is expanding in many countries.

APPENDIX I

Ionising radiation and how it is measured

The following are four kinds of nuclear radiation:

Alpha particles

These are particles (helium nuclei) consisting of two protons and two neutrons and are emitted from naturally-occurring heavy elements such as uranium and radium, as well as from some man-made transuranic elements. They are intensely ionising but can be readily stopped by a few centimetres of air, a sheet of paper, or the human skin. They are only dangerous to people if they are inhaled or ingested and released inside the body. Alpha-radioactive substances are safe if kept in any sealed container, even a plastic bag.

Beta particles

These are either electrons or positrons (therefore of very low mass) emitted by many radioactive elements. They can be stopped by a few millimetres of wood or aluminium. They can penetrate a little way into human flesh but are generally less dangerous to people than gamma radiation. Exposure produces an effect like sunburn, but which is slower to heal. Beta-radioactive substances are also safe if kept in appropriate sealed containers.

Gamma rays

These are high-energy beams almost identical to X-rays and of shorter wavelength than ultraviolet radiation. They are emitted in many radioactive alpha and beta decays. They are very penetrating, and need substantial thicknesses of heavy materials such as lead, steel or concrete to shield them. Gamma rays are the main hazard to people dealing with sealed radioactive materials used, for example, in industrial gauges and radiotherapy machines. Doses can be detected by the small badges worn by workers handling any radioactive materials. Gamma activity in a substance (e.g. rock) can be measured with a scintillometer or Geiger counter.

Neutrons

These are mostly released by nuclear fission, and apart from a little cosmic radiation they are seldom encountered outside the core of a nuclear reactor. Fast neutrons are very penetrating as well as (indirectly) being strongly ionizing and hence very destructive to human tissue. They can be slowed down (or 'moderated') by wood, plastic, or (more commonly) by graphite or water.

X-rays are also ionising radiation, virtually identical to gamma rays, but not nuclear in origin.

Cosmic radiation consists of very energetic particles, mostly protons, which bombard the Earth from outer space.

Alpha, beta, gamma radiation and X-rays do not cause the body or any other material to become radioactive.

Units

The amount of ionising radiation absorbed in tissue can be expressed in grays: 1 Gy = 1 J/kg. However, since neutrons and alpha particles cause more damage per gray than gamma or beta radiation, another unit, the sievert (Sv), is used in setting radiological protection standards. One gray of beta or gamma radiation has one sievert of biological effect, one gray of alpha particles has a 20 Sv effect and one gray of neutrons is equivalent to around 10 Sv (depending on their energy).

Total dose is thus measured in sieverts, millisieverts (mSv) *i.e.* one-thousandth of a sievert, or microsieverts (μSv) – one-millionth of a sievert. The rate of dose is measured in milli- or microsieverts per hour or year. The average natural dose for humans is around 2 mSv/yr from environmental sources. In industry, the maximum annual dose allowed for radiation workers is 20 mSv/yr above natural levels, averaged over five years; in practice, doses are usually kept well below this level.

These levels contrast with those which are known to be harmful to humans: with gamma radiation a short-term dose of 1 Sv causes (temporary) radiation sickness; 5 Sv would kill about half the people within a month of receiving it; a burst of 10 Sv would be fatal to all in a matter of days. The 28 radiation fatalities who died within four months of the Chernobyl disaster appear to have received more than 5 Sv in a few days, while those who suffered acute radiation sickness but recovered averaged doses of 3.4 Sv.

The becquerel (Bq) is a unit or measure of actual radioactivity in material, as distinct from the radiation it emits, or the human dose from that. It indicates the number of nuclear disintegrations per second (1 Bq = 1 disintegration/sec). Quantities of radioactive material are commonly estimated by measuring the amount of intrinsic radioactivity in becquerels – one Bq of radioactive material is that amount which has an average of one disintegration per second i.e. an activity of 1 Bq. This is a very small unit, so gigabecquerels (GBq) etc are more commonly used.

Health effects from radiation are of two kinds: acute or deterministic are effects caused by large doses in a short period, and chronic, or stochastic lead to an increased risk of cancer from low doses over a long period. Acute effects may be skin burns or vomiting.

Older units of radiation measurement continue in use in some literature:

    1 gray = 100 rads
    1 sievert = 100 rem
    1 becquerel = 27 picocuries or 2.7×10^{-11} curies

One curie was originally the activity of one gram of radium-226, and represents 3.7×10^{10} disintegrations per second (Bq).

Radon and radon progeny

The Working Level Month (WLM) has been used as a measure of dose for exposure to radon and in particular, radon decay products (see Appendix 2). One 'Working Level' is approximately equivalent to 3700 Bq/m^3 of Rn-222 in equilibrium with its decay products. Exposure to 0.4 WL was the maximum permissible for workers. Continuous exposure during working hours to 0.4 WL would result in a dose of 5 WLM over a full year[1], corresponding to about 50 mSv/yr whole body dose for a 40-hour week. In mines, individual workers' doses are kept below 1 WLM/yr (10 mSv/yr), and typically average half this.

A background radon level of 40 Bq/m^3 indoors and 6 Bq/m^3 outdoors, assuming an indoor occupancy of 80%, is equivalent to a dose rate of 1 mSv/yr and is the average for most of the world's inhabitants. The World Health Organisation in 1996 set 1000 Bq/m^3 as the reference level for exposure, but in 2009 lowered it to 100 Bq/m^3 indoors as a desirable target, saying that 3 to 14% of cancers are attributable to radon.

Some comparative radiation doses:

2 mSv/year:	Typical background radiation dose rate in Australia.
3 mSv/year:	Typical background radiation to North American public.
3-5 mSv/year:	Typical occupational dose rate (above background) to uranium miners in Canada and Australia.
10 mSv/year:	Maximum actual dose rate to Australian uranium miners.
20 mSv/year:	Current limit for nuclear industry employees (five year average).
50 mSv/year:	Former long-term limit for nuclear industry employees and uranium miners; current maximum limit in single year.
350 mSv in lifetime:	Criterion for relocating people after Chernobyl accident.
1000 mSv as short-term dose:	Likely to cause (temporary) radiation sickness.
10,000 mSv as short-term whole-body dose:	Fatal within a few weeks.

[1] Assuming 170 working hours per month.

APPENDIX 2
Some radioactive decay series showing half-lives

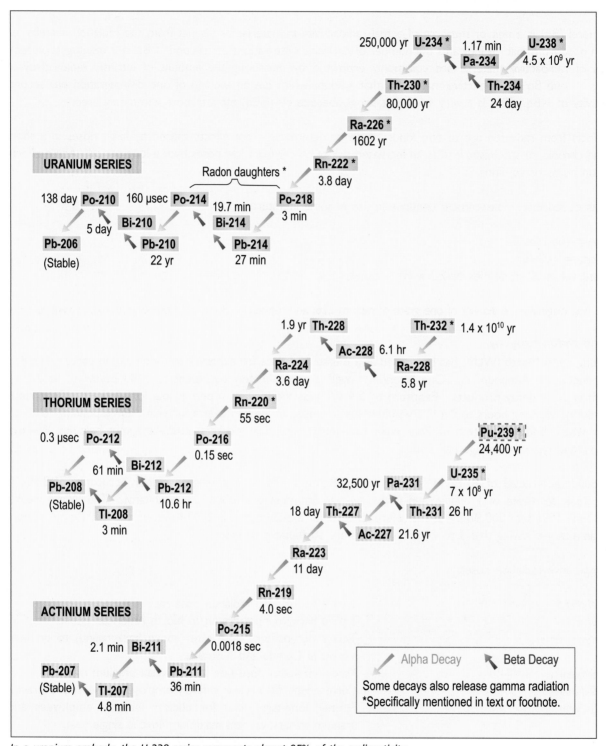

In a uranium orebody, the U-238 series represents almost 95% of the radioactivity.
The level of radiation emitted by an isotope is inversely proportional to its half-life. The shorter the half-life of an isotope, the more radiation it emits per unit mass. Th-232, U-235 and U-238 are thus virtually stable.

APPENDIX 3

Scientific consensus on ethical, societal and technical aspects of radioactive waste management

Several statements published since 1995 serve to underline the scientific consensus that has been reached on the ethical, societal and technical aspects of radioactive waste management.

The first statement was formulated and published by the International Atomic Energy Agency (IAEA) in 1995 to confront the question of what are the best and most appropriate means of managing and disposing of radioactive wastes from the civil nuclear fuel cycle. The IAEA statement (below) has since become part of the *Joint Convention on the Safety of Spent Fuel Management and on the Safety of Radioactive Waste Management*[1], which entered into force in 2001.

The IAEA view was reinforced by a 1995 report issued by the OECD's Nuclear Energy Agency (NEA), titled *The Environmental and Ethical Basis of the Geological Disposal of Long-Lived Radioactive Waste – A Collective Opinion of the Radioactive Waste Management Committee*[2]. The numerous experts from diverse countries involved in preparing this considered that "from an ethical standpoint, including long-term safety considerations, our responsibilities to future generations are better discharged by a strategy of final disposal than by reliance on stores which require surveillance, bequeath long-term responsibilities of care, and may in due course be neglected by future societies whose structural stability should not be presumed." This opinion was endorsed by the IAEA and the European Commission. A 1999 OECD/NEA statement[3] updated the consideration of geological disposal policies.

The OECD Nuclear Energy Agency's 1995 *Collective Opinion of the Radioactive Waste Management Committee*, and the 1999 statement *Progress Towards Geologic Disposal of Radioactive Waste*, are summarised in the WNA information paper on *Waste Management in the Nuclear Fuel Cycle, Appendix 5 – Environmental and Ethical Aspects*[4].

International Atomic Energy Agency
Fundamental Principles of Radioactive Waste Management

A 1995 publication within the IAEA's Radioactive Waste Safety Standards (RADWASS) program defines the objective of radioactive waste management and the associated set of internationally agreed principles. The principles set out in the document are:

1-5. Radioactive waste shall be managed in such a way:
- as to secure an acceptable level of protection for human health.
- as to provide an acceptable level of protection of the environment.
- as to assure that possible effects on human health and the environment beyond national borders will be taken into account.
- that predicted impacts on the health of future generations will not be greater than relevant levels of impact that are acceptable today.
- that will not impose undue burdens on future generations.
6. Radioactive waste shall be managed within an appropriate national legal framework including clear allocation of responsibilities and provision for independent regulatory functions.
7. Generation of radioactive waste shall be kept to the minimum practicable.
8. Interdependencies among all steps in radioactive waste generation and management shall be appropriately taken into account.
9. Safety of facilities for radioactive waste management shall be appropriately assured during their lifetime.

---

[1] http://www.iaea.org/Publications/Documents/Conventions/jointconv.html
[2] http://www.nea.fr/html/rwm/reports/1995/geodisp/geological-disposal.pdf
[3] *Progress Towards Geologic Disposal of Radioactive Waste: Where Do We Stand? An International Assessment*, http://www.nea.fr/html/rwm/reports/1999/progress.pdf
[4] http://www.world-nuclear.org/info/Environmental_Ethical_Aspects_inf04ap5.html

APPENDIX 4
Some useful references

World Energy Outlook 2011
OECD International Energy Agency
Paris 2011
ISBN 978-92-64-12414-1

Uranium 2009: Resources, Production and Demand ('Red Book')
Joint report by the OECD Nuclear Energy Agency and the International Atomic Energy Agency
Paris 2010
ISBN 978-92-64-04789-1

Radioactive Waste Management in Perspective
OECD Nuclear Energy Agency
Paris 1996
ISBN 92-64-14692-X

Radiation in Perspective, applications, risks and protection
OECD Nuclear Energy Agency
Paris 1997
ISBN 92-64-15483-3

Chernobyl's Legacy: Health, Environmental & Socio-economic Impacts
The Chernobyl Forum (eight UN agencies and three governments)
IAEA Vienna 2006

Invisible Rays: A History of Radioactivity
G.I. Brown,
Sutton Publishing, Stroud 2002
ISBN 978-07-50-92667-6

Radiation and Modern Life: fulfilling Marie Curie's dream
Alan Waltar,
Prometheus Books, New York 2004
ISBN 978-15-91-02250-3

All About Nuclear Energy: from Atom to Zirconium
Bertrand Barre
Areva 2008

GLOSSARY

The following is a list of terms which are commonly used in discussion of the nuclear fuel cycle.

Actinide: an element with atomic number of 89 (actinium) to 103. Usually applied to those above uranium – 93 up (also called transuranics). Actinides are radioactive and typically have long half-lives. They are therefore significant in wastes arising from nuclear fission, e.g. in used fuel. They are fissionable in a fast reactor. Minor actinides are americium, curium and neptunium.

Activation product: A radioactive isotope of an element (e.g. in the steel of a reactor core) which has been created by neutron bombardment.

Activity: the number of disintegrations per unit time inside a radioactive source. Expressed in becquerels.

ALARA: As Low As Reasonably Achievable, economic and social factors being taken into account. This is the optimisation principle of radiation protection.

Alpha particle: A positively-charged particle emitted from the nucleus of an atom during radioactive decay. Alpha particles are helium nuclei, with 2 protons and 2 neutrons.

Atom: A particle of matter which cannot be broken up by chemical means. Atoms have a nucleus consisting of positively-charged protons and uncharged neutrons of similar mass. The positive charges on the protons are balanced by the same number of negatively-charged electrons in motion around the nucleus.

Background radiation: The naturally-occurring ionising radiation which every person is exposed to, arising from the Earth's crust (including radon) and from cosmic radiation.

Barn: See *Neutron cross-section*.

Base-load: That part of electricity demand which is continuous, and does not vary over a 24-hour period. Approximately equivalent to the minimum daily load.

Becquerel: The SI unit of intrinsic radioactivity in a material. One Bq indicates one disintegration per second and is thus the activity of a quantity of radioactive material which averages one decay per second. (In practice, GBq or TBq are the common units.)

Beta particle: A particle emitted from an atom during radioactive decay. Beta particles are generally electrons (with negative charge) but may be positrons.

Biological shield: A mass of absorbing material (e.g. thick concrete walls) placed around a reactor or radioactive material to reduce the radiation (especially neutrons and gamma rays respectively) to a level safe for humans.

Boiling water reactor (BWR): A common type of light water reactor (LWR), where water is allowed to boil in the core thus generating steam directly in the reactor vessel (cf. *Pressurised water reactor*).

Breed: To form fissile nuclei, usually as a result of neutron capture, possibly followed by radioactive decay.

Breeder reactor: See *Fast breeder reactor* and *Fast neutron reactor*.

Burn: The process of undergoing fission (analogous to burning a fossil fuel) or otherwise becoming denatured in the reactor core.

Burnable poison: A neutron absorber included in the fuel which progressively disappears and compensates for the loss of reactivity as the fuel is consumed. Gadolinium is commonly used.

Burn-up: Measure of thermal energy released by nuclear fuel relative to its mass, typically Gigawatt days per tonne of fuel (GWd/t).

CANDU: CANadian Deuterium Uranium reactor, moderated and cooled by heavy water (except for the ACR design, which is cooled by light water). These are the most common PHWRs (cf. *Heavy water reactor*).

Centrifuge: A cylinder spinning at high speed to physically separate gas components of slightly different mass, e.g. uranium hexafluoride with U-235 and U-238 atoms.

Chain reaction: A reaction that stimulates its own repetition, in particular where the neutrons originating from nuclear fission cause an ongoing series of fission reactions.

Cladding: The metal tubes containing oxide fuel pellets (cf. *Zircaloy*).

Concentrate: See *Uranium oxide concentrate (U₃O₈)*.

Control rods: Devices to absorb neutrons so that the chain reaction in a reactor core may be slowed or stopped by inserting them further, or accelerated by withdrawing them.

Conversion: Chemical process turning uranium oxide (U_3O_8) into UF_6 preparatory to enrichment.

Coolant: The liquid or gas used to transfer heat from the reactor core to the steam generators or directly to the turbines.

Core: The central part of a nuclear reactor containing the fuel elements and any moderator.

Critical mass: The smallest mass of fissile material that will support a self-sustaining chain reaction under specified conditions.

Criticality: Condition of being able to sustain a nuclear chain reaction.

Cross section: See *Neutron cross-section*.

Decay: Disintegration of atomic nuclei resulting in the emission of alpha or beta particles (usually with gamma radiation). Also the exponential decrease in radioactivity of a material as nuclear disintegrations take place and more stable nuclei are formed.

Decommissioning: Removal of a facility (e.g. reactor) from service, also the subsequent actions of safe storage, dismantling and making the site available for other use.

Deconversion: The chemical process of turning uranium hexafluoride (UF_6) into uranium oxide. Typically depleted UF_6 may be processed for long-term storage in a more stable chemical form. Hydrogen fluoride is a by-product.

Delayed neutrons: Neutrons released by fission products up to several seconds after fission. These enable control of the fission in a nuclear reactor.

Depleted uranium: Uranium having less than the natural 0.7% U-235. As a by-product of enrichment in the fuel cycle it generally has 0.25-0.30% U 235, the rest being U-238. Can be blended with highly-enriched uranium (e.g. from weapons) to make reactor fuel.

Deuterium: 'Heavy hydrogen', a stable isotope having one proton and one neutron in the nucleus. It occurs in nature as one atom to 6500 atoms of normal hydrogen. (Hydrogen atoms contain one proton and no neutrons.)

Disintegration: Natural change in the nucleus of a radioactive isotope as particles are emitted (usually with gamma rays), making it a different element (cf. *Decay*).

Electron volt: 1.6×10^{-19} joules, the amount of kinetic energy gained by a single electron when it accelerates through an electrostatic potential difference of one volt.

Element: A chemical substance that cannot be divided into simpler substances by chemical means; atomic species with same number of protons (being the atomic number of the element).

Dose: The energy absorbed by tissue from ionising radiation. One gray is one joule per kg, but this is adjusted for the effect of different kinds of radiation, and thus the sievert is the unit of dose equivalent used in setting exposure standards.

Enriched uranium: Uranium in which the proportion of U-235 (to U-238) has been increased above the natural 0.7%. Reactor-grade uranium is usually enriched to about 3.5% U-235, weapons-grade uranium is more than 90% U-235.

Enrichment: Physical process of increasing the proportion of U-235 to U-238. See also *Separative Work Unit (SWU)*.

Fast breeder reactor (FBR): A fast neutron reactor (q.v.) configured to produce more fissile material than it consumes, using fertile material such as depleted uranium in a blanket around the core.

Fast neutron: A neutron released during fission, travelling at very high velocity (20,000 km/s) and having high energy (c. 2 MeV).

Fast neutron reactor (FNR): A reactor with no moderator and hence utilising fast neutrons. It normally burns plutonium while producing fissile isotopes in fertile material such as depleted uranium (or thorium).

Fertile (of an isotope): Capable of becoming a fissile isotope by capturing a neutron, possibly followed by radioactive decay; e.g. U-238, Pu-240.

Fissile (of an isotope): Capable of capturing a slow

(thermal) neutron and undergoing nuclear fission, e.g. U-235, U-233, Pu-239.

Fissionable (of an isotope): Capable of undergoing fission: if fissile, by slow neutrons; otherwise, by fast neutrons.

Fission: The splitting of a heavy nucleus into two, accompanied by the release of a relatively large amount of energy and usually one or more neutrons. It may be spontaneous but usually is due to a nucleus absorbing a neutron and thus becoming unstable.

Fission products: 'Daughter' nuclei resulting either from the fission of heavy elements such as uranium, or the radioactive decay of those primary daughters. Usually highly radioactive.

Fossil fuel: A fuel based on carbon presumed to be originally from living matter, e.g. coal, oil, gas. Burned with oxygen to yield energy.

Fuel assembly: Structured collection of fuel rods or elements, the unit of fuel in a reactor.

Fuel fabrication: Making reactor fuel assemblies, usually from sintered UO_2 pellets which are inserted into zircaloy tubes, comprising the fuel rods or elements.

Gamma rays: High energy electromagnetic radiation from the atomic nucleus, virtually identical to X-rays.

Genetic mutation: Sudden change in the chromosomal DNA of an individual gene. It may produce inherited changes in descendants. Mutation in some organisms can be made more frequent by irradiation (though this has never been demonstrated in humans).

Giga: One billion units (e.g. one gigawatt is 10^9 watts or one million kW).

Graphite: Crystalline carbon used in very pure form as a moderator, principally in gas-cooled reactors, but also in Soviet-designed RBMK reactors.

Gray (Gy): The SI unit of absorbed radiation dose, one joule per kilogram of tissue.

Greenhouse gases: Radiative gases in the Earth's atmosphere which absorb long-wave heat radiation from the Earth's surface and re-radiate it, thereby warming the Earth. Carbon dioxide, methane and water vapour are the main ones.

Half-life: The period required for half of the atoms of a particular radioactive isotope to decay and become an isotope of another element.

Heavy water: Water containing an elevated concentration of molecules with deuterium ('heavy hydrogen') atoms.

Heavy water reactor (HWR): A reactor which uses heavy water as its moderator, e.g. Canadian CANDU (q.v.) which is a pressurised HWR (PHWR).

High-level waste (HLW): Extremely radioactive fission products and transuranic elements (usually after plutonium is removed for recycling) in used nuclear fuel. It may be separated by reprocessing the used fuel, or the used fuel containing those isotopes may be regarded as high-level waste. HLW requires both shielding and cooling.

High-enriched uranium (HEU): Uranium enriched to 20% U-235 or more. (That in weapons is about 90% U-235.)

In situ leaching (ISL): The recovery by chemical leaching of minerals from porous orebodies without physical excavation. Also known as **in situ recovery (ISR)** or solution mining.

Intermediate-level waste (ILW): Radioactive waste which requires shielding to protect people nearby, but not cooling.

Ion: An atom or molecule that is electrically-charged because of loss or gain of electrons.

Ionising radiation: Radiation (including alpha particles) capable of breaking chemical bonds, thus causing ionisation of the matter through which it passes and damage to living tissue.

Irradiate: Subject material to ionising radiation. Irradiated reactor fuel and components have been subject to neutron irradiation and hence become radioactive themselves.

Isotope: An atomic form of an element having a particular number of neutrons. Different isotopes of an element have the same number of protons but different numbers of neutrons and hence different atomic masses, e.g. U-235, U-238. Some isotopes are unstable and decay (q.v.) to form isotopes of other elements.

Laser enrichment: Uranium enrichment by photo-dissociation of UF_6 to UF_5^+, using tuned laser beams to break the molecular bond of one of the six fluorine atoms connected to a U-235 atom.

Light water: Ordinary water (H_2O) as distinct from heavy water.

Light water reactor (LWR): A common nuclear reactor cooled and usually moderated by ordinary water. It is a generic designation including BWR and PWR types.

Low-enriched uranium (LEU): Uranium enriched to less than 20% U-235. (That in power reactors is usually 3.5-5.0% U-235.)

Low-level waste (LLW): Radioactive waste which can be handled safely without shielding.

Megawatt (MW): A unit of power, 10^6 watts. MWe refers to electric output from a generator, MWt to thermal output from a reactor or heat source (*i.e.* the gross heat output of a reactor itself, typically around three times the MWe figure).

Metal fuels: Fuels made using natural uranium metal, as used in a gas-cooled reactor.

Micro: One-millionth of a unit (e.g. microsievert is 10^{-6} Sv).

Milling: Process by which minerals are extracted from ore, usually at the mine site.

Mixed oxide fuel (MOX): Reactor fuel which consists of both uranium and plutonium oxides, usually about 5% Pu, which is the main fissile component.

Moderator: A material such as light or heavy water or graphite used in a reactor to slow down fast neutrons by collision with lighter nuclei so as to expedite further fission.

Natural uranium: Uranium with an isotopic composition as found in nature, containing 99.3% U-238, 0.7% U-235 and a trace of U-234. Can be used as fuel in heavy water-moderated reactors.

Neutron: An uncharged elementary particle found in the nucleus of every atom except hydrogen. Solitary mobile neutrons travelling at various speeds originate from fission reactions. Slow (thermal) neutrons can in turn readily cause fission in nuclei of 'fissile' isotopes, e.g. U-235, Pu-239, U-233; and fast neutrons can cause fission in nuclei of 'fertile' isotopes such as U-238, Pu-239. Sometimes atomic nuclei simply capture neutrons.

Neutron cross-section: An indication of the probability of an interaction between a particle and a target nucleus, expressed in barns (1 barn = 10^{-24} cm^2).

Nuclear reactor: A device in which a nuclear fission chain reaction occurs under controlled conditions so that the heat yield can be harnessed or the neutron beams utilised. All commercial reactors are thermal reactors, using a moderator to slow down the neutrons.

Nuclide: Elemental matter made up of atoms with identical nuclei, therefore with the same atomic number and the same mass number (equal to the sum of the number of protons and neutrons).

Oxide fuels: Enriched or natural uranium in the form of the oxide UO_2, used in many types of reactor.

Plutonium: A transuranic element, formed in a nuclear reactor by neutron capture. It has several isotopes, some of which are fissile and some of which undergo spontaneous fission, releasing neutrons. Weapons-grade plutonium is produced in special reactors to give >90% Pu-239, reactor-grade plutonium contains about 30% non-fissile isotopes. About one-third of the energy in a light water reactor comes from the fission of Pu-239, and this is the main isotope of value recovered from reprocessing used fuel.

Pressurised water reactor (PWR): The most common type of light water reactor (LWR), it uses water at very high pressure in a primary circuit and steam is formed in a secondary circuit.

Radiation: The emission and propagation of energy by means of electromagnetic waves or particles (cf. *Ionising radiation*).

Radioactivity: The spontaneous decay of an unstable atomic nucleus, giving rise to the emission of radiation.

Radionuclide: A radioactive isotope of an element.

Radiotoxicity: The adverse health effect of a radionuclide due to its radioactivity.

Radium: A radioactive decay product of uranium often

found in uranium ore. It has several radioactive isotopes. Radium-226 decays to radon-222.

Radon (Rn): A heavy radioactive gas given off by rocks containing radium (or thorium). Rn-222 is the main isotope, from decay of radium-226.

Radon daughters: Short-lived decay products of radon-222 (Po-218, Pb-214, Bi-214, Po-214).

Reactor pressure vessel: The main steel vessel of a nuclear reactor containing the reactor fuel, moderator and coolant under pressure.

Repository: A permanent disposal place for radioactive wastes.

Reprocessing: Chemical treatment of used reactor fuel to separate uranium and plutonium and possibly transuranic elements from the small quantity of fission products and transuranic elements, leaving a much reduced quantity of high-level waste (cf. *Waste* and *High-level waste*).

Separative Work Unit (SWU): This is a complex unit which is a function of the amount of uranium processed and the degree to which it is enriched, *i.e.* the extent of increase in the concentration of the U-235 isotope relative to the remainder. The unit is strictly: kilogram Separative Work Unit, and it measures the quantity of separative work (indicative of energy used in enrichment) when feed and product quantities are expressed in kilograms.

For example, to produce one kilogram of uranium enriched to 3.5% U-235 requires 4.3 SWU if the plant is operated at a tails assay 0.30%, or 4.8 SWU if the tails assay is 0.25% (thereby requiring only 7.0 kg instead of 7.8 kg of natural uranium feed).

About 100-120,000 SWU is required to enrich the annual fuel loading for a typical 1000 MWe light water reactor. Enrichment costs are related to electrical energy used. The gaseous diffusion process consumes some 2400 kWh per SWU, while gas centrifuge plants require only about 60 kWh/SWU.

Sievert (Sv): Unit indicating the biological damage caused by radiation dose measured in Gray (q.v.). One Gray of beta or gamma radiation absorbed has 1 Sv of biological effect; 1 Gy of alpha radiation has 20 Sv effect and 1 Gy of neutrons has 10 Sv effect (cf. *Dose*).

Spallation: The abrasion and removal of fragments of a target which is bombarded by protons in an accelerator. The fragments may be protons, neutrons or other light particles.

Spent fuel: Used fuel assemblies removed from a reactor after several years' use and treated as waste. Often it is another term for *Used fuel*.

Stable: Incapable of spontaneous radioactive decay.

Tailings: Ground rock remaining after particular ore minerals (*e.g.* uranium oxides) are extracted.

Tails: Depleted uranium (cf. *Enriched uranium*), with about 0.2 to 0.3% U-235.

Thermal reactor: A reactor in which the fission chain reaction is sustained primarily by slow neutrons, and hence requiring a moderator (as distinct from a fast neutron reactor).

Transmutation: Changing atoms of one element into those of another by neutron bombardment, causing neutron capture and/or fission. In an ordinary reactor neutron capture is the main event; in a fast reactor fission is more common and therefore it is best for dealing with actinides. Fission product transmutation is by neutron capture.

Transuranic element: A very heavy element formed artificially by neutron capture and possibly subsequent beta decay(s). Has a higher atomic number than uranium (92). All are radioactive. Neptunium, plutonium, americium and curium are the best-known.

Uranium (U): A mildly radioactive element with two isotopes which are fissile (U-235 and U-233) and two which are fertile (U-238 and U-234). Uranium is the basic fuel of nuclear energy.

Uranium hexafluoride (UF_6): A compound of uranium which is a gas above 56°C and is thus a suitable form in which to enrich the uranium.

Uranium oxide concentrate (U_3O_8): The mixture of uranium oxides produced after milling uranium ore from a mine. Sometimes loosely called 'yellowcake'. It is khaki in colour and is usually represented by the empirical formula U_3O_8. Uranium is sold in this form.

Used fuel: Fuel assemblies removed from a reactor after several years' use.

Vitrification: The incorporation of high-level wastes into borosilicate glass, to make up about 14% of it by mass. It is designed to immobilise radionuclides in an insoluble matrix ready for disposal.

Waste: High-level waste (HLW) is highly radioactive material arising from nuclear fission. It can be what is left over from reprocessing used fuel, though some countries regard used fuel itself as HLW. It requires very careful handling, storage and disposal.

Intermediate-level waste (ILW) comprises a range of materials from reprocessing and decommissioning. It is sufficiently radioactive to require shielding and is disposed of in engineered facilities underground.

Low-level waste (LLW) is mildly radioactive material usually disposed of by compaction or incineration and then burial.

Yellowcake: Ammonium diuranate, the penultimate uranium compound in U_3O_8 production, but the form in which mine product was sold until about 1970. See also *Uranium oxide concentrate.*

Zircaloy: Zirconium alloy used as a tube to contain uranium oxide fuel pellets in a fuel rod (part of a reactor fuel assembly).

INDEX